Human-Computer Interaction Series

Editors-in-chief

John Karat
Jean Vanderdonckt, Université Catholique de Louvain, Belgium

Editorial Board

Gaëlle Calvary, LIG-University of Grenoble 1, France
John Carroll, Penn State University, USA.
Gilbert Cockton, Northumbria University, Newcastle, UK
Larry Constantine, University of Madeira, Portugal
Steven Feiner, Columbia University, USA
Peter Forbrig, Universität Rostock, Germany
Elizabeth Furtado, University of Fortaleza, Brazil
Hans Gellersen, Lancaster University, UK
Robert Jacob, Tufts University, USA
Hilary Johnson, University of Bath, UK
Dianne Murray, Putting People Before Computers, UK
Kumiyo Nakakoji, University of Tokyo, Japan
Philippe Palanque, Université Paul Sabatier, France
Oscar Pastor, University of Valencia, Spain
Fabio Pianesi, Istituto Trentino di Cultura, Italy
Costin Pribeanu, National Institute for Research & Development in Informatics, Romania
Gerd Szwillus, Universität Paderborn, Germany
Manfred Tscheligi, Center for Usability Research and Engineering, Austria
Gerrit van der Veer, Vrije Universiteit Amsterdam, The Netherlands
Shumin Zhai, IBM Almaden Research Center, USA
Thomas Ziegert, SAP Research CEC Darmstadt, Germany

Human-computer interaction is a multidisciplinary field focused on human aspects of the development of computer technology. As computer-based technology becomes increasingly pervasive – not just in developed countries, but worldwide – the need to take a human-centered approach in the design and development of this technology becomes ever more important. For roughly 30 years now, researchers and practitioners in computational and behavioral sciences have worked to identify theory and practice that influences the direction of these technologies, and this diverse work makes up the field of human-computer interaction. Broadly speaking it includes the study of what technology might be able to do for people and how people might interact with the technology.

In this series we present work which advances the science and technology of developing systems which are both effective and satisfying for people in a wide variety of contexts. The Human-Computer Interaction series will focus on theoretical perspectives (such as formal approaches drawn from a variety of behavioral sciences), practical approaches (such as the techniques for effectively integrating user needs in system development), and social issues (such as the determinants of utility, usability and acceptability).

For further volumes:
www.springer.com/series/6033

Jürgen Steimle

Pen-and-Paper User Interfaces

Integrating Printed and Digital Documents

Forewords by
James D. Hollan
and
Max Mühlhäuser

Jürgen Steimle
Department of Computer Science
Technische Universität Darmstadt
Darmstadt, Germany

Foreword by
James D. Hollan
Department of Cognitive Science
University of California, San Diego
La Jolla, CA, USA

Foreword by
Max Mühlhäuser
Department of Computer Science
Technische Universität Darmstadt
Darmstadt, Germany

Many of the designations used by manufacturers and sellers to distinguish their products are claimed as trademarks. Where those designations appear in this book, and Springer was aware of a trademark claim, the designations have been printed in initial capital letters. However, not all words in initial capital letters are trademark designations.

While every precaution has been taken in the preparation of this book, the publisher and the author assume no responsibility for errors or omissions, or for damages resulting from the use of the information contained herein.

ISSN 1571-5035 Human-Computer Interaction Series
ISBN 978-3-642-20275-9   e-ISBN 978-3-642-20276-6
DOI 10.1007/978-3-642-20276-6
Springer Heidelberg Dordrecht London New York

Library of Congress Control Number: 2011944662

ACM Classification (1998): H5, H5.2

© Springer-Verlag Berlin Heidelberg 2012

This work is subject to copyright. All rights are reserved, whether the whole or part of the material is concerned, specifically the rights of translation, reprinting, reuse of illustrations, recitation, broadcasting, reproduction on microfilm or in any other way, and storage in data banks. Duplication of this publication or parts thereof is permitted only under the provisions of the German Copyright Law of September 9, 1965, in its current version, and permission for use must always be obtained from Springer. Violations are liable to prosecution under the German Copyright Law.

The use of general descriptive names, registered names, trademarks, etc. in this publication does not imply, even in the absence of a specific statement, that such names are exempt from the relevant protective laws and regulations and therefore free for general use.

Printed on acid-free paper

Springer is part of Springer Science+Business Media (www.springer.com)

Foreword

Computers provide the most plastic medium for representation, communication, and interaction we have ever known. The computational medium is plastic in the sense that we can employ it to:

- mimic other media (e.g., books, newspapers, magazines, photographs, audio recordings, and films), devices, and mechanisms of interaction,
- create models that represent, with ever increasing fidelity, the physical world, spanning from models of atoms and molecules to those used to forecast weather or to guide spacecraft to destinations far from earth,
- provide virtual worlds that range from the simple metaphorical desktop of the graphical user interface to the amazing digital effects and virtual characters of current films, or
- combine the real and the virtual (e.g., in robotic surgery the tremors of a surgeons's hands are removed as he or she interacts with a computer interface remote from the patient).

This plasticity and the myriad ways computers are now enmeshed in our personal and professional lives and in the infrastructure of science and society present enormous opportunities and challenges. Computationally-based forms of communication and interaction are changing the world in which we live and the ways we interact within it.

With each new technology we seem characteristically drawn to focus almost exclusively on the new opportunities it presents and how it might replace older technologies, frequently forgetting, or at least not adequately appreciating, that new technologies must exist and evolve in ecologies comprised of older technologies as well as webs of established cognitive and cultural practices. Understanding how a new technology meshes and interacts with existing ecologies and practices is as fundamental and important a component of design as the new opportunities it creates.

From the beginnings of the modern computer at Xerox Parc in the early '70s one consistent refrain has been that new computational devices would replace paper and the future would be increasingly paperless. As every office and home continues to bear witness, and as is well documented in Sellen and Harper's seminal book, The Myth of the Paperless Office, we are far from becoming paperless, continuing to exploit both digital and paper media. Following Wellner's early digital desk explorations there have been a series of innovative investigators exploring not how to replace paper with digital tools but rather how to combine the two. Stellar examples of this approach are Guimbretière's Paper Augmented Digital Documents, Yeh's Butterflynet, Liao's PapierCraft, Tabard's hybrid notebook, and Signer's and Weibel's work to support interaction across the paper-digital divide. The most recent advance in this line of research is the excellent thesis work of Jürgen Steimle that is the basis of the present book. Steimle confronts the complex and crucially important issue of how to bridge and combine digital and paper worlds so as to facilitate access to the best of each.

In this book Steimle addresses the question of how to design interfaces that integrate traditional pen-and-paper-based practices with digital media. He begins with a comprehensive survey of pen-and-paper computing in which he covers the technologies involved, existing toolkits and applications, and characterizes an underlying model of generic core interactions as a basis for developing principles and guidelines. Based on this model he describes CoScribe, a novel modular framework to support collaborative paper-based work. CoScribe provides an integrated environment that supports multiperson collaboration with multiple documents. It is unique in addressing the complex challenges involved in supporting asynchronous shared handwritten annotations and hyperlinking between printed and digital documents. The exposition using scenarios, detailed descriptions of the technologies, and careful empirical evaluations is compelling and advances both the science and technology of interface design. While the focus of the book is on pen-and-paper interfaces, everyone interested in how to design for real-world activity will profit from reading this book.

San Diego, October 2011 *James D. Hollan*

Distributed Cognition and Human Computer Interaction Lab
Department of Cognitive Science
University of California, San Diego

Foreword

The mass deployment of Smart Phones, Netbooks, and Web tablets has made computing essentially pervasive and ubiquitous – yet to date, there is only one truly ubiquitous information processing technology: pen and paper. Imagine the wealth of paper variants that may populate the venue of a creative and information-centered workshop: little Post-it notes and snippets, sturdy colored cards, numerous piles of memos and notes, groupings of bound or stapled documents, annotated leaflets and brochures, journals and magazines along with ready-made, commented laser copies of relevant contents, binders and folders full of classified information, flipcharts and wall-covering series of charts and other drawings, not to mention waste baskets, full to the brim with torn and crumpled sheets, ... the list is endless. And it is still a long way to go until we can use and afford computers in the same quantity and variety, and with such simplicity and carelessness.

On the other hand, computers can handle information in a way that paper will never be able to: store and archive in 'infinite' quantity with an ever smaller footprint, search and analyze, transmit, copy, and share at virtually no cost at lightning speed, edit and interconnect, ... again, the list is endless. Given these considerations about the uniqueness and ubiquity of both paper and computers, the present book is long since overdue: a thorough, concise, and well-organized compendium of marriages between paper based and electronic documents.

This book, a revised and extended version of Jürgen Steimle's award-winning computer science dissertation, provides the reader with a broad and extensive overview of the field. The state of the art is covered in a most up to date, complete, and systematic way, so as to provide the full picture of pen-and-paper computing like no other reference before.

The contributions made with regard to modeling the interaction with pen-and-paper interfaces provide an unprecedented theoretical foundation and organization of the subject matter, helping to structure and order the problem and design spaces in a rather unique way. The book proposes information ecologies as the appropriate theoretical perspective for designing pen-and-paper interfaces. This involves taking a broad view and looking at all the ingredients that largely influence the interplay of humans and machines in the context of information handling: current and related documents, cognitive and social networks, past actions and future-oriented intentions. The author presents an elegant 'building set' of core interactions helpful in designing solutions that address the diversity of such ecologies.

Retaining the holistic approach of the book, the third part presents an integrated set of interaction techniques for the most relevant human document processing activities: collaboratively annotating, combining (linking), and classifying (tagging) documents. Here, the aforementioned systematic theoretical framework forms the basis for the cleanest and most flexible approach known in comparison to related work. Regarding cross-media annotation, the presented approach provides an impressive proof of the huge potential that lies in joining the individual strengths of the two technologies, paper and computing. As to combination i.e. hyperlinks, a rather small advancement in hardware is provided as a basis: the enabling of Anoto technology for use with both computer screens and traditional paper. This small technical contribution enables a huge effect with respect to eliminating seams and hurdles between the two technologies. Finally, concerning classification (tagging), the author provides smart and elegant means for tagging documents with predefined classes, but also with arbitrary tags that are defined on-the-fly. Here and in the aforementioned contributions, the author proves to be quite resourceful when it comes to leveraging the strengths of paper as a technology, such as the flexible interplay of many paper sheets, but also when it comes to coping with its limitations, such as the lack of inverse operations for writing or cutting.

In short, the present book promises to be an exciting source of information for IT professionals (trying to understand the cutting-edge field of pen-and-paper computing), researchers (interested in an overview of prior research and in the substantial original academic contributions presented in this book), and HCI experts (seeking insights into the comparatively young field of pen-and-paper computing as well as on the advancement of their field in general).

Darmstadt, October 2011 *Max Mühlhäuser*

Telecooperation Lab
Department of Computer Science
Darmstadt University of Technology

Acknowledgements

This publication would not have been possible without the support and encouragement of many persons. I wish to acknowledge their contribution.

First and foremost, I acknowledge and thank Max Mühlhäuser (TU Darmstadt), the primary advisor of my doctoral dissertation, for his unlimited support and excellent advice. I am also highly grateful to Jim Hollan (UC San Diego), Jan Borchers (RWTH Aachen) and Werner Sesink (TU Darmstadt) for fruitful discussions and for their valuable feedback on earlier versions of this book.

Many thanks are also due to all present and former members of the Telecooperation lab at TU Darmstadt. They provided a very friendly place to work and supported this work in innumerable respects. In particular I would like to thank the undergraduate and graduate students who contributed to the implementation and evaluation of the prototype system: Stefan Buhrmester, Roman Lissermann, Simon Olberding, Michael Stieler, Sasa Vukancic and Jie Zhou.

Support for my research has been generously provided by a grant from the DFG – German Research Foundation. As a research fellow of the DFG postgraduate school "E-Learning", I profited enormously from the highly inspiring atmosphere and the cross-disciplinary discussions. In particular, I wish to thank Oliver Brdiczka (now at Palo Alto Research Center) and Christoph Koenig for their advice.

I gratefully acknowledge Scott Klemmer, Andreas Paepcke and Ron Yeh (Stanford University) for their invaluable advice and feedback on my work during my lab visit. Moreover I address special thanks to the many researchers who supported this publication by providing photos of their own research: Florian Block, Raimund Dachselt, Katherine Everitt, Michael Haller, Scott Hudson, Scott Klemmer, Chunyuan Liao, Peter Ljungstrand, Wendy Mackay, Beat Signer, Hyunyoung Song, Nadir Weibel, Pierre Wellner and Andy Wilson.

Last but not least I am grateful to Olga Chiarcos and her colleagues from Springer for their excellent assistance in the production of this manuscript as well as to the external reviewers for their valuable feedback.

Contents

1 Introduction 1
   1.1 Why Using Paper Documents? 3
   1.2 Integrating Pen, Paper, and Computers 9
   1.3 Challenges 13
   1.4 Contributions and Structure of this Book 14
   1.5 How to Read this Book 17

2 Survey of Pen-and-Paper Computing 19
   2.1 Technologies 19
      2.1.1 Digitizing Contents of Paper Documents 19
      2.1.2 Page Identification and Location Tracking 21
      2.1.3 Capturing Touch Input 26
      2.1.4 Capturing Pen Input 27
      2.1.5 Digital Output on Paper 33
      2.1.6 Pen-and-Paper Toolkits 35
   2.2 Pen-and-Paper Interfaces 39
      2.2.1 Augmented Paper Cards and Post-Its 40
      2.2.2 Augmented Books 41
      2.2.3 Augmented Paper Notebooks 42
      2.2.4 Augmented Printed Documents 48
      2.2.5 Augmented Tables, Flipcharts and Whiteboards 58
   2.3 Directions of Future Research 64

3 Interaction Model of Pen-and-Paper User Interfaces 67
   3.1 Pen-and-Paper User Interfaces (PPUIs) 69
   3.2 Related Models 71
   3.3 An Ecological Perspective of Document Work 74
   3.4 Model of Interactions 77
      3.4.1 Semantic Level of Interaction: Conceptual Activities .. 78
      3.4.2 Syntactic Level of Interaction: Core Interactions 79
      3.4.3 Mapping Between Syntax and Semantics 83

		3.5	Model of Information 85
		3.6	Conclusions and Design Guidelines 88
4	**CoScribe: A Platform for Paper-based Knowledge Work** 91		
	4.1	Main Conceptual Activities 93	
	4.2	Interaction Tools 94	
	4.3	Synchronized Paper Documents and Digital Visualizations 95	
	4.4	Collaboration 99	
	4.5	Implementation 101	
5	**Collaborative Cross-media Annotation of Documents** 103		
	5.1	An Adaptable Printed User Interface for Annotations............. 104	
	5.2	Paper-based Sharing of Annotations 108	
	5.3	Visualization of Shared Annotations 111	
	5.4	Evaluation and Discussion 115	
		5.4.1	Study I: Field Study of Lecture Annotation................ 115
		5.4.2	Study II: Laboratory Study of Annotation Review 119
		5.4.3	Study III: Performance of Handwriting Recognition 122
6	**Hyperlinking between Printed and Digital Documents** 127		
	6.1	Unified Pen-based Linking on Paper and on Displays............. 129	
	6.2	Creating and Following Hyperlinks............................. 131	
	6.3	Sharing of Hyperlinks... 138	
	6.4	Ecological View... 138	
	6.5	Evaluation and Discussion 141	
7	**Paper-based Tagging of Documents** 149		
	7.1	Tangible Tagging with Stickers: Digital Paper Bookmarks 151	
	7.2	Tagging by Association: Tag Menu Card 157	
	7.3	Tagging with Buttons 159	
	7.4	Tangible Tagging of Processes................................. 159	
	7.5	Evaluation and Discussion 162	
8	**Conclusions** 167		
	8.1	Summary of this Book 167	
	8.2	Directions of Future Research 172	

References 177

Index 187

Chapter 1
Introduction

> *The paperless office is a myth (...) because (people) know (...) that their goals cannot be achieved without paper. This held true over thirty years ago when the idea of the paperless office first gained some prominence, and it holds true today at the start of the twenty-first century. (...) It will hold true for many years to come.*
>
> A. Sellen and R. Harper, The Myth of the Paperless Office

Paper has been used over thousands of years. Even though digital media are getting increasingly more sophisticated, paper is surprisingly persistent. Paper is certainly used differently than some decades ago. However, even at the beginning of the twenty-first century, it is still pervasive in our homes, workplaces, schools and universities.

During the last decades, many attempts aimed at replacing paper documents by digital media. Desktop computing, word processing, electronic mail and the World Wide Web have been considered to have a large potential for replacing paper. In contrast, paper was considered a symbol of old-fashioned technology. However, the numerous predictions of the paperless office have not become reality [131].

Where does this omnipresence of paper stem from? Why cannot paper be easily replaced by computer technology? The longevity of paper cannot be merely attributed to shortcomings of current display technology, such as limited screen size, resolution and contrast. A large body of research shows that paper supports a wealth of interactions that have a number of inherent advantages over digital technologies. To state only some of these advantages, annotating paper documents with a pen is easy, flexible and smoothly integrated with reading. In addition, paper renders information tangible. People can utilize their both hands for interacting with it and get tactile-kinesthetic feedback. This provides for effectively navigating within a document, for example when thumbing through a book and sensing the appropriate number of remaining pages with one finger, but also for sorting and structuring paper-based information. These are only some examples of the advantages of paper documents, which we will identify in more detail below.

Despite these advantages of paper, it is a matter of course that digital media have other, equally important benefits. For instance, digital documents can be efficiently searched, archived and shared over a distance. Moreover, they can include dynamic contents, including audiovisual and interactive media. In addition, albeit it is relatively inexpensive to produce paper and to print on it, the cost of dealing with paper documents after printing – delivery, storage and retrieval – can be much higher than the respective cost of digital documents [131].

Due to the unique benefits of both worlds outlined above, people typically use not only paper or only digital documents. Rather they *combine* both worlds. Depending on the type of information and the context of use, some information is preferred in a printed form while other information is accessed using digital technology. For example, paper might be preferred for reading a longer document, while a computer might be the tool of choice for composing new documents or for looking up information on the Web. This combined use of printed and digital documents leads to disruptive transitions. Users must cope with different representational media as well as with different interactions and tools. Most important, while many digital documents can be easily printed on paper, the reverse direction is more challenging. It is still difficult to efficiently digitize paper-based information.

During the past two decades, a new area of research has formed that develops technical solutions for the integration of paper-based and digital information. Rather than replacing one medium by the other, the main goal of this strand of research is to reduce the gap between printed and digital documents and to combine the best of both worlds. Since many paper-based activities also involve using pens, most interfaces do not only support using physical paper, but also physical pens, and make both of them key elements of digital user interfaces. This presents novel opportunities for improving computer support for document-based activities. Research on Pen-and-Paper User Interfaces can be assigned to the fields of Ubiquitous Computing and Tangible User Interfaces, which aim at extending computing beyond the computer desktop into the physical space that surrounds us.

The present books inscribes into this strand of research and focuses on how to integrate pens, physical paper, and computers. The main question addressed in this book is as follows:

How to design user interfaces that effectively integrate traditional pen-and-paper-based practices with digital documents?

Our answer is three-fold. Each part of the book addresses this overall question from a different perspective:

First, this book provides a comprehensive overview of prior research on Pen-and-Paper User Interfaces. At the time of publication of this book, this is the most complete and up-to-date survey of the field. It gives extensive insights into technologies, technical frameworks and existing concepts for user interfaces.

Second, looking at Pen-and-Paper User Interfaces on a rather abstract, conceptual level, the book introduces a generic interaction model. Going beyond the individual interaction techniques presented in prior work, this model provides systematic guidelines for designing Pen-and-Paper User Interfaces.

Third, the book provides a concrete instantiation of the model: the CoScribe framework. CoScribe introduces an integrated set of interaction techniques that support effective knowledge work[1] with documents. We introduce novel concepts that support collaboration on various levels, integrate paper and screens closer than before, and make ample use of the physical flexibility of paper. All these techniques are generic and can be easily integrated into user interfaces that target different work settings.

This introductory chapter serves for framing the topic of this book. In order to understand what is so specific about paper, we first provide a synopsis of prior research that has examined the affordances of paper. Next we will present our approach – integrating printed with digital documents – and introduce the basics of Pen-and-Paper Interfaces. This allows us to outline key challenges that will be addressed in this book. Finally, we provide an overview of the chapters and guide readers through the book's structure.

1.1 Why Using Paper Documents?

With the advent of word processing, electronic mail and the World Wide Web, many experts predicted that the end of paper use in offices was imminent. Entire companies attempted going paperless, thereby banishing the symbol of old-fashioned technology. However the paperless office has failed to materialize. A number of similar predictions of paperlessness can be traced back in the history until the 19th century [131]. Paper survived each of them.

Recently, novel technologies, such as e-book readers and tablets, have come to the market. Printed newspapers see themselves challenged by online information portals. A growing number of scientific works get published solely online. Again many commentators prognosticate that paper is becoming passé. Indeed, electronic media are currently pushing back paper to some extent in publishing. While in the 1990s, paper production was constantly increasing, paper consumption currently seems to have reached a plateau and remains approximately at the level of the year 2000 [13].

However, this does not mean that we will go paperless to work. The key point is that the publication medium is not necessarily the medium that we use for eventually working with the document. For instance, Sellen and Harper showed that the electronic access to documents made possible by the World Wide Web did not reduce but it even increased paper consumption. One reason is that people prefer reading long documents on paper [131]. This still holds true for state-of-the art e-readers and tablet devices [153, 103]. Hence, even at the beginning of the twenty-first century, paper remains a key information medium that is omnipresent in our homes and at our workplaces.

In this section, we discuss what are the affordances of paper that make it such a pervasive medium, despite all advances in technology. What affordances are likely

---

[1] Following Drucker [25], we define *knowledge work* as a category of work which primarily deals with using and developing information. Some very obvious examples of knowledge workers are: scientists, teachers, students, librarians, engineers, lawyers, journalists.

to be equally provided by computer technology in the near future? What characteristics of paper are unlikely to be successfully addressed by computer technology in the foreseeable future? To answer these questions we will provide a brief survey of main results from the literature that studied workplace practices of using paper.

In their seminal work *The Myth of the Paperless Office* [131], Sellen and Harper identified main activities that paper supports in knowledge work: Paper supports reviewing documents, amongst others because paper documents can be flexibly annotated and commented. Although it might appear counterintuitive, paper is also a key part in authoring processes. Even though people typically use a word processor for composing documents, authoring also consists of reading, planning and thinking, which is facilitated by paper. Finally, the use of paper documents proves to be supportive for collaboration and organizational communication. Reading is a key element of all these activities.

We systematize the main reasons why paper supports knowledge work so successfully along six key aspects. The survey shows that with respect to display quality and mobile use, the advance of paper is diminishing. However, paper offers specific advantages that current technology does not provide in this form. This includes easy navigation, intuitive annotation, flexible organization of information in the physical space as well as strong support of collaboration and mutual awareness.

Image Quality

The quality of the image which is presented to the reader is undoubtedly an important issue for reading [23]. High resolution, high contrast, little flicker, and a sufficiently large size of the display all are crucial aspects. Only a few years ago, the comparison of image quality between a good paper print and a good computer screen clearly resulted in favor of paper. Empirical studies showed that reading from screens was slower than reading from paper (even though this did not necessarily negatively affect comprehension rates, at least with short texts) [23].

In the meanwhile, display technology has dramatically evolved, and today's state-of-the-art screens provide much better image quality than some years ago. On the one hand, the screen real estate which is available at a usual office workplace has significantly increased. Nowadays it is not unusual to have a screen as large as 27 inch or two smaller screens that feature an even larger overall size. These screens can display not only a single entire page of a document, but even multiple pages simultaneously in a readable size. This display size might even be too large and overwhelming for some users, as reported by Morris et al. [99]. On the other hand, resolution and contrast are constantly increasing. Many displays provide a resolution of 150 dpi or more. Displays of modern smart phones provide the resolution of even more than 300 dpi, which is comparable to print resolution. It is only a matter of time when larger screens start featuring similar resolution. Moreover, novel display technologies, such as e-paper and OLED, allow for very high contrast, similarly to paper.

Hence, screens are currently measuring up to paper in terms of image quality. While the reading experience of e-books may still be worse than the reading ex-

1.1 Why Using Paper Documents? 5

perience of books on paper,indexE-book reader in the near future, image quality is likely to cease being a unique asset of printed information. This leads to an assumption that in the future, reading will be done more frequently on screens than today.

However, a large body of research shows that image quality is not the only factor that is crucial for reading and other knowledge work processes. For instance, Dillan [23] summarizes a large number of empirical studies and concludes that the most obvious difference between reading from paper and reading from screen is the ease with which paper can be manipulated. In the remainder of this section, we will address these affordances of paper, which go beyond the issue of image quality.

Navigation in Documents

People read documents only rarely straight from the beginning to the end. Often they read a document only partially or they need to make connections between different parts of a a documents. Navigation in documents is therefore an intrinsic part of the reading process. By its physical nature, paper helps us easily and flexibly navigate through documents [131, 23, 110]. Empirical results show that navigation through a paper document is more implicit and more tightly interwoven with reading than navigation through digital documents. Due to the tactile nature of interaction with paper, many activities can be carried out with little to no visual attention. For instance, when turning pages the thickness of a document is used as a physical cue for estimating its length and the remaining number of pages. Moreover, people make heavy use of both hands for searching and skimming through the document. This allows for interleaving navigation with other activities, for instance flipping pages while writing with the other hand.

Due to the static layout of paper documents, information is fix with respect to a physical page. O'Hara et al. [110] show that users acquire incidental knowledge of the location of information by reference to its physical place on the page, which helps them find this information later on. In this respect, digital document formats that feature a fix layout (such as PDF) are preferable to formats that reflow text to fit the current width of the window (such as most Web pages do).

Handwritten Annotations

Opposed to what we intuitively understand by reading – deciphering words and phrases and ultimately meaning – writing is a substantial element of reading processes. Sellen and Harper [131, p. 82] point out that in knowledge work, reading occurs with writing more often than it occurs without. In a diary study, the participants combined reading with writing (taking notes or making annotations[2]) in more

---

[2] We define an annotation as an amendment to an existing document that is conceptually separate yet contextually related. It adds an additional layer of information to it leaving the original document unchanged. In contrast to an annotation, a note has none or only a weak contextual relationship to an existing document. Most often, annotations are made on top of an existing document, while notes are taken on a separate, initially empty sheet of paper. We will not always clearly dis-

than 75 % and up to 91 % of the time. Commenting, underlining and highlighting a document during reading supports better understanding, critical thinking as well as remembering the thoughts the reader had. Adler et al. call this process *Active reading* [2]. As Adler notes, "the physical act of writing, with your own hand, brings words and sentences more sharply before your mind and preserves them better in your memory". Annotations and notes are not only central to reading but also important for efficiently attending meetings or lectures. Psychological research shows that notetaking plays an important role in learning processes and has been proven to be a factor positively related to students' academic achievement [58, 115].

An important affordance of paper documents is that they can be easily annotated. Handwritten annotations can consist of lines and brackets that highlight portions of the document. They can also contain text, formulae, sketches, etc. This shows their great fluidity in form [92]. They can be very informal or more structured, for example if the user follows a specific annotation or notetaking method (e.g. the Cornell Notetaking Method [113]). Moreover, handwritten annotations on a printed document are clearly separated from the document. Finally, several persons can annotate the same copy of a document [131].

People make most of their annotations on paper, even when the document is available in a digital version [131, 153]. The results of field studies that we have conducted to inform our work show that university students clearly preferred printed versions of lecture scripts to their digital counterparts, mainly for ease of annotation [145]. The results also show that using a pen and paper handouts in a seminar results in a higher number of annotations than using a tool for typewritten annotations on a laptop [143, p.28 sqq.]. Similar in approach, Obendorf [109] compared paper-based annotation with annotations made on Web pages. He found that in the paper condition much more annotations were made.

Yet, these results do not imply that handwritten annotations are in all cases preferable to typewritten ones. Pen and paper are clearly beneficial for quick and implicit annotations that are tightly interwoven with reading. In contrast, typewritten annotations might be advantageous as soon as the annotation is made less implicitly, is longer and formulated with more care. Think for instance of formulating a summary of a book chapter. The point is that in the latter case, the annotation process can be considered as related rather to composition than to reading.

Pen-enabled displays emulate many of the characteristics of pen and paper. This poses the question whether they are as good as paper for making handwritten annotations. Empirical studies come to differing results. There is some empirical evidence that the use of pen-enabled displays generates greater extraneous cognitive load than the more familiar interactions with real paper [112]. Morris et al. [99] report qualitative findings indicating that pen-enabled horizontal and tablet displays provide an annotation experience largely comparable to pen and paper. However, the participants were still suffering from difficulties stemming from the annotation software, such as the overhead of entering a special inking mode. Piper and Hollan [116] compared handwritten annotations on paper and on horizontal displays that were

---

tinguish between annotating and notetaking because on a technical level, handwritten notes can be modeled as annotations on an empty document.

made during group discussions. They report that the more ephemeral character of annotations in the digital condition encouraged participants more than with paper to spontaneously create annotations to support discussion. In contrast, participants of the digital condition often erased their annotations, which hindered them in going back and reviewing them for subsequent reflection. Moreover participants who used paper made more detailed notes. Subsequent research is necessary to compare annotations on paper and on pen-enabled displays in more detail.

Spatial Organization of Information

Organizing and structuring information are crucial activities in knowledge work. To state only two examples, this comprises making connections between different documents or different parts of a document as well as organizing and prioritizing information during planning and thinking activities. The resulting structural knowledge facilitates recall and comprehension and is essential to problem solving [51]. The use of physical space turns out to be a very effective support for such activities.

Paper affords cross-page and cross-document use. By freely arranging several pages on the desk, readers dispose of multiple "display" surfaces. This physical arrangement enables to lay out information in space. For instance, placing several sheets of a document one besides the other facilitates getting an overview of the document. Physical arrangements also ease comparing or integrating information from several documents [110], which is a very frequent practice [1]. People intuitively and dynamically adapt the physical arrangement of paper sheets accounting for the activity which is currently performed. For example, reading requires placing the paper sheet at a different angle and at a different distance from the reader than writing [110]. Finally, people make use of their both hands to seamlessly interweaving multiple activities. For instance, one document can be used for reading, while another one is placed besides it for taking notes.

The physical arrangement of documents also provides rich ways of expressing functions and priorities as well as relationships between them. For instance, creating a pile of documents is a lightweight and informal strategy of organizing information on the desk [90]. While piling does not scale well to long-term archiving of a large number of documents, empirical research has shown that piling leads to a more frequent access to information when compared to more formal strategies of organization [169].

The literature shows the sophisticated spatial patterns that people apply implicitly for organizing space on the desk. Sellen and Harper point out that the typical workplace of a knowledge worker contains different functional zones for documents of different priorities [131]. They distinguish between "hot", "warm" and "cold" zones. The hot zone contains documents that are within the user's center of attention. Warm zones contain documents of lower priority, which should not interfere with the current task but nevertheless be quickly at hand. Warm zones are typically located at the outer zones of the desk. Cold zones contain documents that are currently of little or no relevance. For instance these documents are filed in shelves in the same or even in another room. In collaborative settings, Scott et al. found similar, yet even more flexible patterns of organizing space [129]. They identified personal territories, group territories and storage territories, which are used for placing documents that are of little relevance for the current task. These zones are not clearly delimited; rather they form a continuum with smooth transitions. They can grow larger and get smaller over time. Storage territories are mobile and move over the table while the task progresses.

Summing up, spatial arrangements are important for expressing and gaining a sense of the overall structure, for classifying documents and for referring to other documents. The question is whether these spatial arrangements are specific for paper or whether large horizontal displays, so-called interactive tabletops, can provide the same advantages. Interactive tabletops [98] use the surface of tables as an interactive display. The screen is typically large enough to display several documents side-by-side. Direct touch interactions allow users to freely arrange digital documents on the screen. A limiting factor of most current tabletops is their small resolution. The XGA or HD resolution, which is typically available, in combination with the large display size, results in a rather small pixel density. While this is acceptable for images, text is not readable in a size comparable to that of printed text. The future will solve this issue. More important are aspects related to the haptic characteristics of paper. Interweaving of activities, such as flipping through the pages of one document for skimming and simultaneously writing on another document, is obviously more difficult without haptic feedback. Terrenghi et al. [151] compared the manipulation of physical versus digital objects on table surfaces. They found that even if digital tabletop interfaces copy many of the characteristics of traditional tables, the resulting interaction is fundamentally different. With digital tabletops there is little of the ease with which we manipulate physical objects. For instance, participants required the double average amount of time for solving a puzzle task with digital objects than when they used physical objects. Moreover, participants perceived the interaction with physical objects as more convenient. Results of one of our own studies [146] show that physical documents are used not only on the tabletop surface, but to a large extent also above or in front of the surface, even when they are used in conjunction with digital documents. Main reasons are first that documents are held at an angle which is more comparable for reading than when placed flat on the surface. Second, documents can be easily compared and sorted when they are held in the user's hands. In summary, large interactive tabletops provide enough space for flexibly arranging documents; however, current solutions lack the important haptic aspects of paper.

Mobile Use

A further key affordance of paper is mobile use. Since paper is thin, flexible and lightweight, pen and paper can be easily taken along and used in a huge variety of situations and physical places. At least this is true for short documents. If it comes to carrying several heavy books or an entire archive, paper is not very mobile, of course. As a further benefit, pen and paper is always-on, since no battery back must be charged. Even in environments where only expensive specific computer hardware can be utilized (e.g. extreme temperatures, humidity, dust), pen and paper is working. Finally, paper is cheap. Hence, in most cases it is less problematic if a paper document gets damaged, lost or stolen than a more expensive computer device.

Paper is not only mobile at this "macro-level", where paper is carried to different physical places. Luff and Heath showed how important the micro-mobility of paper is [84]. By micro-mobility, they refer to the many small movements that we make with a paper document which is at hand. Since most paper documents are rather lightweight, we can easily reconfigure them during an interaction to best serve the current activity. For instance, the document is held at different angles to support comfortable reading depending on whether one is leaning back or leaning over the table. It is also hold differently depending on the activity (e.g. reading vs. writing) and depending on the communicative situation.

Collaboration

Finally, paper has specific affordances that support collaboration and awareness of the activities of co-workers. First, using paper documents leaves implicit and explicit traces [91, 92], which are helpful for subsequent readers [175]. For example, a textbook in a library, which has been used for some years, contains implicit traces of use. Nagged and stained pages indicate passages that have been read by many borrowers. Annotations or dog-ears made by previous readers are more explicit traces of use. Marshall [91] states that it is precisely for these traces that many students prefer buying second-hand textbooks instead of new ones.

Paper also supports mutual awareness in co-located collaborative settings. At a single glance, even from a peripheral viewpoint, it is easy to see if a person works with little or many documents and if she is reading or writing. Mackay [87] clearly demonstrates how these features effectively support collaboration in air traffic control.

Finally, there are subtle social aspects of how paper supports communication. Even though it is typically more effort to personally hand over a document to a co-worker than sending an electronic mail, people frequently prefer the physical way. Studies show that physical hand-over of documents stimulates personal communication [87] and can even be a means for reconfirming the social order by the specific way the document is handed over [105].

1.2 Integrating Pen, Paper, and Computers

It is not surprising that the vision of the paperless office has not become true. Even though state-of-the-art computing devices, such as interactive tabletops, e-book readers and tablet computers, have significantly improved, there still remain too many of the specific advantages of paper that they do not provide. Most important are its tangible characteristics, which allow us to navigate through documents with ease, to efficiently organize information in space, and to effectively communicate.

Motivated by the workplace studies from the literature and a number of own studies in which we analyzed how university students use paper and digital media [145, 144, 143, p. 19 sqq.], we are strongly convinced that rather than replacing paper, combining paper with computers is more promising. This combines the ease and flexibility of using paper with the powerful capabilities of computing. It is left to the user to choose the adequate medium for a given task. Several technical approaches allow for combining paper with digital documents:

Printing and Scanning A first approach consists of using two well-established tools for bridging between both worlds: printers and scanners. If a printed version of a digital document is required, the user prints it. In the reverse direction, if a digital version of some paper document is required, the document is scanned. These tools are quite powerful for one-way transformations. However, this is different once the user wants to perform a cyclic transformation. Imagine that the user prints a document on paper, makes handwritten annotations, and then wants to integrate these annotations back into the digital version. Such a process is difficult and time-consuming to perform. Moreover, scanning has a low update rate and therefore does not allow more interactive uses of paper that have immediate effects on the digital side.

Interactive Paper Another class of approaches extends the user interface of a computer system to paper. Thereby physical paper becomes interactive. Interacting on or with paper does not only alter the physical paper sheet, but also controls the computer system. Interactive paper is often understood as synonymous to using a digital pen on physical paper. The digital pen behaves like a traditional pen and leaves visible, physical ink traces on paper. In addition, the pen captures the traces electronically and sends them to a computer system. The traces can be visualized as a facsimile of the handwritten content or can be interpreted as commands for controlling the computer system. In addition to pen input, some interactive paper systems also identify whether paper documents are present and how they are arranged to react accordingly on the digital side. For instance, a camera tracks the location of paper documents on a desk and captures their contents. A projector projects additional digital information onto the same surface [167]. As another example, the detailed arrangement of several paper documents on an interactive tabletop controls how digital information is laid out on the surface [57].

Electronic Paper Whereas interactive paper solutions use real paper, electronic paper [16] is a novel display technology. It provides for very thin, bendable and lightweight displays. Even though these displays are not paper, they share many of its physical characteristics. Currently the development of electronic paper is in a rather early stage. Electronic paper displays available on the market are still rigid and quite thick. Hence, many of the affordances of paper that we have identified in the previous section are not reached yet. While it can be expected that this will change in the future, there will be many years to come until electronic paper can be used like real paper. In this book, we therefore focus on interactive paper solutions that combine traditional paper with computers.

Pen-and-Paper User Interfaces (PPUI)

In this book, we investigate the most widespread class of interactive paper solutions, which is Pen-and-Paper User Interfaces (PPUI). A PPUI minimally consists of real paper and a digital pen. Usually a further computing device is part of the interfaces, such as a PC, a laptop, an interactive tabletop or a smart phone. An example is depicted in Fig. 1.1.

What is specific about PPUIs is that paper becomes a user interface. One or several sheets of paper contain printed user interface elements. The main interaction device is a digital pen whose position on the paper sheet is automatically tracked. This allows the user to enter digital data by writing and/or drawing on the paper sheet. Moreover, the user can enter additional interactional information, for instance issuing a command by performing a pen "click" on a virtual button which is printed on paper. The system provides digital feedback either via the pen itself using a built-in display, LEDs, audio or haptic feedback. Alternatively it prints an updated version of the paper sheet(s) and/or provides feedback using nearby devices, such as smart phones, standard computer displays or projectors.

Fig. 1.1 Example of a Pen-and-Paper User Interface

Fig. 1.2 A digital pen (Anoto ADP-301, copyright Anoto)

PPUIs have the advantage that the technology is available on the market, affordable and can be easily integrated into established work practices. Moreover, writing with a pen is one of the most important forms of working with documents. Digital pens retain the flexibility of free-form input, which is central to knowledge acquisition as we have discussed above. PPUIs can be realized with several different pen technologies. Each of them has different advantages and drawbacks. We will discuss these technologies in the following chapter. The currently most advanced technology is Anoto Digital Pen and Paper [4]. Figure 1.2 depicts an Anoto pen.

In the example shown above in Fig. 1.1, a PDF document is printed on several paper sheets. All interactions that the user makes with the pen on these paper sheets are automatically captured and interpreted by the computer system. On the one hand, the digital pen can be used for annotating the document. Handwritten annotations that are made on the printed version of the document are automatically integrated into the digital document. On the other hand, each sheet of paper contains additional user interface elements. Several interactive areas behave like virtual buttons. When the user taps with the pen on one of these buttons, a command is triggered. For instance, the buttons allow the user to categorize annotations with a label and share them with co-workers. Feedback is given directly on the pen and on a nearby computer display. A digital version of the document can be accessed on the display, including handwritten annotations and their categories. If necessary, the user can also print an updated version of the document.

A frequent use case of PPUIs consists of annotating printed versions of digital documents, as outlined in the example above. PPUIs are also used to augment traditional paper notebooks with digital functionality. While the paper notebook retains many of the advantages of paper (e.g. mobile use, intuitive handwriting), a digital version is automatically created that includes additional functionality (e.g. digital images can be directly integrated into the notebook). As a further sample use case, PPUIs allow us to design paper-based controls for computer systems. For instance, pen and paper can be used as a remote control for a TV set. These are only some applications of PPUIs. The following chapter will provide a comprehensive overview of concepts and systems.

1.3 Challenges

The novel type of user interface that is made possible by digital pens creates considerable challenges for interface design, since established paradigms for Graphical User Interfaces (GUI) do not, or only partially, apply to this setting. This book addresses the following challenges:

Appropriate Paper-based Interaction Techniques What are appropriate interaction techniques for user interfaces that are printed on paper? Is it possible and desirable to transfer established concepts from Graphical User Interfaces to PPUIs? For instance, GUI dialogs including widgets such as buttons, checkboxes and text input fields could be printed on paper. The pen could replace the mouse. Or is it a more promising approach to take inspiration from the well-established practices of working with paper? How could interaction techniques look like that take inspiration from paper-based interactions? And finally, how can we make sure that the novel interaction techniques fit into how we work with paper *and* how we work with computers?

Restricted Feedback Capabilities Graphical User Interfaces allow the system to provide extensive feedback to the user in real-time. In contrast, PPUIs have a restricted feedback loop. Ideally, PPUIs could update the printed user interface in real-time, similarly to how GUIs can update the user interface on the screen. However, current digital pens cannot print information on paper. Therefore, it would be necessary to reprint the entire sheet of paper for updating information. So real-time feedback, for instance about the system state, has to be given by other channels. Current digital pens can provide only restricted feedback – most pens cannot provide any system feedback at all, and even the most advanced pens feature only a very small display. Therefore, external devices, such as nearby screens, are typically used for this purpose. Yet, the key affordance of pen and paper is its mobility and the fact that no other devices or tools are necessary during mobile use. To retain this key affordance, interactions in a PPUI shall be reliable and clear even if only very limited digital feedback can be provided.

Static vs. Dynamic User Interface This challenge is related to the previous one. Content which is printed on a sheet of paper is static when compared to dynamically updated information which is displayed on a screen. Hence a printed user interface risks to be rather static. Nevertheless, the PPUI should be dynamic and flexibly adapt to the current needs of the user.

Symbiotic Integration into Existing Practices PPUIs should symbiotically integrate into existing and well-established work practices. One challenge is connected to the implicit and highly personal way we use traditional pen and paper. As discussed above, this comprises for instance the way how documents are marked up during reading and the way how collections of documents are organized. It is therefore crucial that digital support is capable of adapting to these personal practices, preferences and habits. A further challenge is connected to the high degree of cognitive load [15] which characterizes many situations of knowledge work. For instance, reading and understanding complicated subjects matters are cognitively demanding tasks. It might be by no means a simple coincidence that traditional tools and interactions for working with documents are characterized by a high degree of simplicity and intuitiveness. Like traditional tools, novel interactions should produce little extraneous load in order to be easily integrated into existing practices.

Collaboration Support As the analysis of the affordances of paper has shown, paper is a collaborative medium. Using paper is very powerful during co-located interaction because its physical nature affords flexible multi-user interaction in co-presence. For example, it is very easy to jointly write on a document. Documents can also be flexibly moved and spatially organized to structure them or to (re-)attribute specific documents to specific persons. However, it is cumbersome to use paper for collaboration over distance. PPUIs offer a benefit here, since interactions with physical documents can be electronically tracked. The question is what are appropriate interactions and visualizations to support paper-based collaboration over a distance.

1.4 Contributions and Structure of this Book

This book is a substantially revised and extended version of the author's doctoral dissertation. As outlined above, our overarching goal is to provide answers to the question of how to design user interfaces that effectively integrate traditional pen-and-paper-based practices with digital documents. Our application domain is knowledge work and we particularly focus on knowledge acquisition tasks. We take on an integrated viewpoint that takes into account the various activities that are central to knowledge work. The scientific contributions that are discussed in this book are situated in the fields of interaction theory, interaction techniques and interactive systems. They comprise a comprehensive survey of related work (Chapter 2), a theoretical interaction model (Chapter 3), and a set of novel interaction techniques which instantiate the model (Chapters 4-7).

Chapter 2 provides an extensive survey of prior research that has been conducted in the field of Pen-and-Paper User Interfaces. It discusses technologies which enable to build bridges between paper documents and the digital realm. Moreover, it gives an overview of technical frameworks that support developing end-user applications with pen and paper. Furthermore, it discusses existing interactive paper systems, interaction concepts and applications.

Chapter 3 introduces a theoretical model of Pen-and-Paper User Interfaces (PPUIs). Prior research in this field has focused on systems and not on theory. The model

1.4 Contributions and Structure of this Book

contributes at filling this gap by sharpening the theoretical understanding of digital interaction with pen and paper. On the one hand it supports the analysis of existing user interfaces. On the other hand it provides systematic guidelines for the design of novel interfaces that are easy to use, reliable and that can be seamlessly integrated into existing work practices.

In three initial field studies, we have explored how paper and digital media are used in knowledge acquisition processes. Based on the results of these studies, we propose an ecological perspective as an appropriate theoretical position for the design of interactive paper systems. This perspective is based on Distributed Cognition and on Information Ecologies and situates interaction within a collaborative work context. This perspective is the basis for the actual interaction model. The model systematically separates interaction into two levels: While the semantic level models what the user wants to perform with the interface, the syntactic level models how this is achieved. This separation enables us to identify a set of generic interaction primitives that can be performed with digital pen and paper. It will be demonstrated that systems from related work can be classified in terms of these interaction primitives. Moreover, these interaction primitives serve as building blocks of Pen-and-Paper User Interfaces. By combining several interaction primitives, complex paper-based user interfaces can be designed that offer a rich user experience while remaining being both reliable and easy to use. This theoretical approach is at the foundation of the novel interaction techniques, which are presented in the following chapters.

Chapters 4–7 present the CoScribe concept for cross-media knowledge work with documents. CoScribe is based on the theoretical interaction model of PPUIs and proposes an integrated solution. Following an ecological model of knowledge work as our guiding theory, CoScribe covers entire workflows in knowledge work and puts strong emphasis on collaboration. Several persons can work at the same place using multiple pens or can collaborate over a network connection.

Chapter 4 provides a high-level overview of the CoScribe concept. CoScribe comprises a set of novel interaction techniques and visualizations for cross-media knowledge work with documents. They focus on the cohabitation [32] of paper and computers, i.e. both media are used in combination and treated at an equal footage. These techniques and visualizations support several activities that are central for effective knowledge acquisition. They enable users to annotate printed and digital documents. Moreover, they allow users to generate structural knowledge of how concepts of a domain are interrelated. This can be done by integrating documents with hyperlinks and by translating contents into higher-level concepts with tags. In terms of functionality, this is similar to Bush's vision of Memex [12], a machine that supports reading and learning processes which are based on documents. In contrast to Memex, the user cannot only work with documents which are displayed by a machine but also with printed documents. The interaction techniques offer a rich user experience, being inspired by the traditional practices of using paper and relying on such varied interactions as writing on paper, arranging several sheets of paper, connecting paper sheets and attaching physical stickers. This varied user interface stands in contrast to many previous approaches that leverage only the interaction of writing with a pen on paper. In addition, the interaction with paper and digital information is more seamlessly integrated than in previous Pen-and-Paper User Interfaces, since the same digital pen and the same interactions can be used both on printed and on digital documents. Nevertheless, both flexibility and mobility of paper are retained.

Chapter 5 presents interaction techniques for paper-based annotation of documents. Taking notes and making annotations is an important activity in Active Reading processes [1] and has been proven to be a factor positively related to students' academic achievement [58, 115]. We introduce a paper-based mechanism for classifying and sharing annotations, which is seamlessly integrated with annotating and notetaking. For the review of shared annotations of other users, we propose a multi-user visualization that integrates annotations of multiple users in one single view. Handwriting recognition is a vital aspect of annotation functionality in order to provide full-text search within handwritten contents. We discuss results from an evaluation study of handwriting recognition which shows that particularly domain-specific terms are hard to recognize. We present an algorithm to augment recognition performance for domain-specific terms in annotations. Finally, we present results from two studies in which the novel interaction techniques have been evaluated with users. They show that the new interaction techniques are easy to learn, easy to use and reliable.

Chapter 6 presents a novel interaction technique for easily and quickly creating and following hyperlinks between paper and digital documents. Creating references between documents is an important means for integrating information and constructing knowledge. While this is a common practice in paper-based settings, and hyperlinks are common on the Web, functionality is missing for easy creation of references between paper and digital documents (e.g. between a book and a Web page). The underlying interaction metaphor of our technique is a pen-based association which crosses the boundaries of paper and screens. We show how pen gestures for linking documents can be designed in order to be easy to learn and to memorize, reliable and efficient. We further outline how hyperlinks can be used in a collaborative setup to support joint structuring activities in workgroups. To cope with the problem known as "lost in hyperspace", we present a multi-user, multi-document visualization that gives an overview of all users, documents and hyperlinks. Finally, we present results of a controlled experiment that evaluates the efficiency of this interaction technique. They demonstrate a significant performance gain for an information integration task in hybrid collections of printed and digital documents.

Chapter 7 presents novel techniques for tagging paper documents. In addition to structuring documents with hyperlinks, tagging enables people to further integrate new knowledge with existing knowledge by relating pieces of information and translating them into concepts. This generates structural knowledge. The knowledge of these relations and the ability to explain them is existential for higher-order procedural knowledge [51]. Our techniques go beyond the previous work on paper-based tagging that is exclusively based on the usage of pen gestures. In contrast, we leverage the flexibility of physical paper sheets, which, for instance, can have different shapes, or can be attached to one another. First, we show how the haptic aspect of paper can be leveraged for tagging documents using physical index stickers. This integrates a well-established, highly efficient traditional paper practice with digital support enabled by a computer interface. Next, we demonstrate further techniques which imply the use of separate cards for tagging and which make use of printed buttons. This underscores the variety of interaction styles and the richness that is possible in paper-based interaction. Then we show that not only documents, but also collaborative processes can be tagged using tangible tools, and discuss how this novel type of tagging can support learning and knowledge work. Finally, we compare the concepts and present the results of user studies.

Chapter 8 summarizes the contents of the book. We then provide an outlook and identify opportunities and directions for future research. These include further exploring the use of paper documents with a range of interactive devices, such as smart phones and electronic paper. Electronic paper is a highly promising technology that has the potential to significantly alter the way we work both with traditional paper and with information on displays. Future work should examine novel interaction techniques for contents on electronic paper. It should also assess how electronic paper can be integrated into existing information ecologies. Mobile access to information through interactive paper documents requires investigating and exploiting a class of interactions that only the combination of two portable and pervasive media like pen-and-paper and mobile phones can support. Moreover, it is still unclear how digital pen-and-paper applications integrate into large-scale social media, such as Web 2.0 applications and social networking. We envision that a new range of innovative lightweight interactions could emerge in these settings. In general, the field is lacking a deeper understanding of the long-term effects of pen-and-paper applications. Therefore, here is a clear need for long-term studies.

1.5 How to Read this Book

This book is written for researchers, students, HCI professionals and IT professionals who are interested in Pen-and-Paper User Interfaces. We suggest the following routes through the book depending on the background and interest of the reader:

Anyone who wants to get a background of the field

Read chapters 1–2 and 8.2

Researchers and HCI professionals interested in conceptual advances of the field

Read Chapter 3. Optionally read Chapters 4–7 for concrete instantiations of the concept

Anyone interested in novel computer support for knowledge work activities

Read Chapters 4–7

HCI professionals who seek guidance in designing Pen-and-Paper User Interfaces

In case you are new to the field, read Chapter 2 first. If you prefer a top-down approach, continue with Chapter 3 for getting conceptual insights before you read Chapters 4–7 and 2.2 for concrete examples of interfaces. If you prefer a bottom-up approach, start with the examples before reading Chapter 3.

HCI and IT professionals who want to decide whether Pen-and-Paper is an appropriate solution for their use case

Read Chapters 1–2

Those who think that there is no need for integrating paper and computers

Read Chapter 1.1 and then decide how to continue your reading

Chapter 2
Survey of Pen-and-Paper Computing

Over several decades, a large body of research has been established that focuses on Pen-and-Paper Computing. This chapter reviews previous work of the field – both from a technological and interface perspective – and discusses future directions of research and development.

2.1 Technologies

Pen-and-Paper Interfaces require that the computer system be able to capture how a user is interacting with physical documents in the real world. This capturing should be robust and perform in real-time without requiring a complicated technological setup or adding interactional overhead. Ideally, paper documents and interactions on paper should be tracked without the need to modify them in any form. In this section, we first discuss technologies for realizing input to a computer system: this includes capturing contents of paper documents, identifying documents and tracking their locations. We further present technologies that allow for capturing input that users make on paper using their hands, fingers or pens. Finally, we briefly discuss how computer output can be provided on paper, by using projection or by augmenting paper with electronic components.

2.1.1 Digitizing Contents of Paper Documents

Even though increasingly more documents are available in an electronic form, there are still documents that only exist on paper. Visual scanners allow us to digitize their contents. A scanner captures a static image of a page's contents at a given point in time. If an image contains text, optical character recognition (OCR) [119] can be used to convert the graphical marks into a machine-readable symbolic representation.

Desktop scanners are well-suited for scanning large numbers of pages and offer a high resolution. Modern desktop scanners feature an automatic document feeder and scan up to 35–60 pages per minute with a resolution of 600 dots per inch or more.

In contrast to desktop scanners, handheld scanners are small and light, so they can be used in mobile settings. A first class of handheld scanners has a scanning unit integrated into the tip of a pen-like device. The interaction for scanning resembles to using a pen on a paper document. By moving the pen along the lines of text, the document is successively scanned. A computing unit built into the pen integrates the small fragments to one single image of the page. This approach is well-suited for scanning individual words or short passages but is too slow for scanning entire pages. An example is the Wizcom InfoScan2 Elite pen[1]. The pen is able to scan with a speed of about 15 cm/s and a resolution of 400 dpi. It features optical character recognition. This enables the pen to read the scanned text aloud the scanned text using voice synthetization, to translate scanned text and to provide definitions for scanned words. Users can transfer scanned data to a computer or a mobile phone via an infrared port or via USB. A second class of handheld scanners is not used like a pen, but is placed flat on the document, very much like a ruler. The scanning module therefore has a larger width, which significantly speeds up the scanning processes. This enables scanning an entire page in 4 to 8 seconds. An example is the Planon DocuPen RC 805[2] (Fig. 2.1).

Fig. 2.1 Planon DocuPen RC 805 handheld scanner (photo copyright Planon Ltd.)

---

[1] http://www.wizcomtech.com (all references to web pages contained within this book were retrieved on 2011-10-10)

[2] http://www.planon.com

2.1 Technologies   21

The advantages of desktop and handheld scanners are that they can scan all types of paper documents without additional provisions made to the documents. However, only one static picture is taken at a given point in time. So, it is not possible to continuously track changes made to the document. Moreover, all visual contents are digitized in one single layer. For example, it is therefore a complex task to separate the underlying printed document from handwritings made on it, such as annotations and sketches. Hence, scanning is not optimal for interactive applications.

Another approach supports more interactive uses. The contents of documents are captured by one or several cameras which are mounted above the table or in front of an interactive wall. While contents can be continuously tracked, camera capturing provides lower resolutions than most desktop and handheld scanners. We will discuss this approach in more detail below.

2.1.2 Page Identification and Location Tracking

Many settings that integrate paper with computing require that a physical sheet of paper can be uniquely identified. For instance, a physical paper card could be used as a physical token to access a specific digital object. Alternatively, the system could help the user to find a paper document in the office by indicating the physical location of the document. In both cases, the system must identify the document. Two main approaches can be distinguished: marker-based and content-based identification. Marker-based approaches require that the objects that are to be tracked contain a machine-readable tag. This tag can be visible to the human eye, such as a printed barcode, or invisibly integrated into the object, such as an electronic RFID tag. Content-based techniques do not interfere with the visible artwork, but result in lower processing speed and can distinguish between only a smaller number of objects than marker-based techniques.

Visual Markers

Visual markers (also called fiducials) encode an identifier in an optical machine-readable representation. A marker is captured by one or several cameras, by a light sensor or by a laser scanner. The most widespread form of visual markers, contained on almost any product, is the linear barcode. It encodes a binary sequence by varying the width of black bars that are arranged in a linear sequence. The EAN 13 coding scheme [44] is used worldwide to identify products at cashpoints. Figure 2.2 (a) shows an example of this barcode.

Linear barcodes are not used by many Pen-and-Paper Interfaces. More common are two-dimensional fiducials, since they allow for storing more data and also for tracking the location of objects in 3d space. Several types of two-dimensional fiducials can be distinguished:

Fig. 2.2 Linear and two-dimensional barcodes ([a]DataGlyph example by Jeff Breidenbach)

Pattern-based fiducials are used for instance in the widely-used open source marker tracking library ARToolkit[3]. The fiducial has a black frame which encloses a black and white pattern (Fig. 2.2 (b)). The library identifies a barcode by comparing the pattern with a set of pre-registered templates using pattern matching techniques from computer vision. The advantage of pattern-based fiducials is that the application developer has some influence on how the fiducial looks like and can create meaningful markers. For instance, it can contain a symbol or some text. However, identification is more error-prone than with other techniques.

Matrix-based fiducials are also known as 2D barcodes. They encode binary data by a two-dimensional grid of black and white points. Each point encodes one bit of data, whereby some bits are usually reserved for error correction. Figure 2.2 (c) depicts an ID marker of ARTag [28]. 2D barcodes allow not only to encode an identifier but a relatively large amount of data. For instance, one single QR code [46] (see Fig. 2.2 d) can store more than 4,000 alphanumeric characters. The DataMatrix [45] code can store more than 2,000 characters. It is for example used for electronic stamps by the German postal service. The data density of 2D barcodes can be further increased by varying visual properties of the points, such as color or brightness, but at the price of a decrease in robustness.

In addition to *identifying* objects, matrix-based fiducials can also be used for tracking the *location* of objects. If the size and shape of the marker is known, it is possible to calculate the marker's relative position and orientation with full 6

---

[3] http://www.hitl.washington.edu/artoolkit/

2.1 Technologies   23

degrees-of-freedom information from a 2D camera image. This principle is widely used in augmented reality and tangible interaction systems. Several toolkits offer out-of-the-box support for application developers. ARTag [28] supports up to 1024 barcode markers which have been optimized for fast and reliable detection. ARToolkitPlus [162] is an improved version of the original ARToolkit[4] that is inspired by ARTag's approach. It offers binary markers that can be more robustly detected than the pattern-based markers of the original ARToolkit.

Topological region adjacency is leveraged by a third class of fiducials. The basic idea of this approach is, instead of encoding a sequence of bits as a sequence of black and white points, to encode a hierarchical graph. The original bit sequence is decoded by identifying and then traversing this graph. The advantage of this approach is that it is fast and very robust against false positive detection [104]. However, only a small number of identifiers can be encoded. reacTIVision [53] is an open-source toolkit that uses this technique. Figure 2.2 (e) shows a reacTIVision fiducial. reacTIVision is used in many projects that require tracking the location and orientation of tangibles on interactive tabletops. While reacTIVision tracks the 2D position and orientation of objects on a flat surface, it cannot provide full 6 degrees-of-freedom information. Recent research showed how to obtain 6 degrees-of-freedom information using topological region adjacency markers [104].

The main advantage of using fiducial markers for identifying and tracking objects is that this is a relatively inexpensive tracking solution. However, this approach requires that the fiducial is in the line-of-sight of the camera. This requirement can severely restrict natural interactions, for instance, when objects are piled. Moreover, the fiducial interferes with the artwork of the object.

Content-embedded fiducials are visually less obtrusive. DataGlyphs [52] (Fig. 2.2 (e)) encodes binary data with a pattern of forward and backward slashes. They are flexible in size, shape and color. This makes it possible to emulate the look of a grayscale or even of a color image by an appropriate pattern of small dashes, similar to how offset printing emulates images by small raster dots. At 600 dpi printing resolution, DataGlyphs can encode up to 1,000 bytes of data per square inch. Another technology, Anoto digital pen and paper [4], performs tracking by decoding a pattern of tiny points that is printed onto paper documents and hardly visible to the human eye. The Anoto approach is presented in more detail in Section 2.1.4 below.

Some applications require real-time location tracking in three dimensions, even in cases when objects are moving very fast. For instance this is important in augmented reality applications that overlay physical objects with projected digital contents. Optical motion capture systems (e.g. Vicon[5], OptiTrack[6]) use several high-speed infrared cameras that observe a scene from different perspectives. Several small retro-reflective dots are attached to the object which is to be captured. These dots appear as white blobs in the camera images. If a dot is seen by at least two

---

[4] http://www.hitl.washington.edu/artoolkit/
[5] http://www.vicon.com
[6] http://www.naturalpoint.com/optitrack

Fig. 2.3 An infrared camera of the OptiTrack high-speed motion capture system

cameras, its 3D position can be calculated. If several dots form a known spatial arrangement, the object's 3D pose can also be identified. Figure 2.3 shows an OptiTrack camera.

Electronic Markers

Electronic markers avoid some of the major problems of visual markers. Radio-Frequency Identification (RFID) uses tags that are embedded into or applied to a physical object. The tag consists of an integrated circuit that stores a unique ID and manages the communication with an external reading device. Moreover, it includes an antenna for receiving and transmitting signals. If the tag is passive, it does not include an own battery but receives energy from the reading device via an electromagnetic field. When the RFID tag is within the range of a reading device, it transmits its unique ID.

No direct line of sight between the tag and the reading device is necessary. However, in contrast to fiducials, RFID tags produce significant costs. To date, an RFID tag still costs more than 0.1 USD. Depending on how paper is used in a Pen-and-Paper Application, these costs might be prohibitive. Moreover, in contrast to fiducials that can be printed directly with the paper document, an additional processing step is necessary for applying RFID tags to paper. Finally, while the technology enables to track whether a tag is within the range of a reader or not, it does not identify its precise location. A novel technology called Near Field Communication (NFC) enables mobile phones to act both as RFID tag and as RFID reading device. NFC is already available in several mobile phones. It is likely that RFID will gain more prominence in the near future and will supersede visual markers in some domains.

A range of electromagnetic solutions (e.g. Ubisense[7]) provide for capturing the physical location of objects. To date, however, they require large markers that cannot be used when working with thin sheets of paper.

Some research projects have developed individual solutions for identifying objects. Similarly to RFID, integrated circuits storing a unique identifier are applied to physical objects. The communication to a reading device is made via a wired connection (e.g. [130, 49]) or a wireless connection (e.g. [114]).

Content-based Identification and Tracking

Content-based approaches do not require markers. Instead, documents are identified and tracked solely by their visual appearance. A camera is observing the scene or documents are scanned, and image processing techniques are used for analyzing the images.

The most commonly used approach is SIFT features [83]. In order to identify a document page, it first has to be registered. The system captures specific optical features of the image and stores these features as a fingerprint of the document page. When a document page appears in the camera image, its features are captured and compared against the fingerprints in the database to identify the page. This technique performs quite well if the camera provides a good resolution and the number of different pages is not very high. For instance, Kim et al. [60] reached a recognition rate of more than 90 % if the document width in the camera image was at least 300 pixels. Liu et al. [79] present an improved set of features that was inspired by the SIFT feature set. They achieved recognition rates of more than 99 %, even if pages had to be identified from a very large set of pages. However, their evaluation excluded lighting effects and camera noise, so it must be assumed that in real setups performance will be lower.

Another approach uses optical character recognition of document text [24]. The recognized text is used for querying a database of documents and retrieving the digital version. As an alternative to using text recognition, the white spaces between words have proven to be quite unique for each text. Brick Wall Coding [26] detects the white gaps between words and takes them as features. It has lower recognition performance than SIFT or FIT features, but it can detect a document page even when the page is only partially visible to the camera.

A drawback of content-based approaches is that the visual appearance must contain enough information for unambiguous identification. For instance, it is not possible to uniquely identify one of several copies of the same document page. Moreover, content-based identification is challenging if the user modifies documents, for instance by making annotations.

---

[7] http://www.ubisense.net

2.1.3 Capturing Touch Input

A further important field for bridging paper and computers is to track touch input on physical sheets of paper. The most frequent approach for detecting touch input on physical paper consists of observing the scene with an overhead camera. Touch events are detected in the camera image using image processing techniques. An influential early example is Wellner's DigitalDesk [167]. The DigitalDesk recognizes the position of fingers by their shape in the video frames. A microphone attached to the bottom of the desk is used to detect the exact moment when the user's finger taps on the desk. However this does not allow the system to detect multiple touches that occur simultaneously. Another technique can detect multiple simultaneous touches. It tracks the color of the fingertips. When the color changes, it is assumed that the finger is pressed against a surface [93]. Further approaches analyze shadows cast by the fingers [171] or use depth information for detecting whether a finger is hovering over or is touching the surface. Wilson [172] recently presented a technique that uses a depth camera to detect touches not only on flat, but also on curved surfaces. Figure 2.4 shows how the approach works: The scene (left) is observed by a depth camera. Initially a model of the surface is set up that stores the depth of the surface for each pixel while no hands or fingers are visible in the camera image. The current input of the depth camera (center picture) is then compared against this model. Hands and fingers have lower depth values because they are situated above the surface. If the distance between the value stored in the model and the currently recorded value falls below a certain threshold, a contact is detected (right).

A drawback of camera-based capturing is that it significantly restricts the mobility of paper. Typically, the camera is mounted at a fix position. While there exist mobile solutions, such as the Docklamp [55] or Sixth Sense [96], these are still rather large and heavy and their use is not comparable to the flexibility of traditional pen and paper. This restriction can be alleviated by using multiple cameras that capture a large volume [173].

Other approaches for detecting touch use electromagnetic field sensing [6] or embed electronics into paper, e.g. push buttons [114, 9].

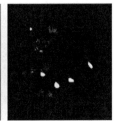

Fig. 2.4 Detecting touch using a depth camera. Left: Touching a book. Center: Depth image. Right: Detected contacts (photos courtesy of Andy Wilson)

2.1 Technologies   27

2.1.4 Capturing Pen Input

Pen-input is an important style of interaction with paper documents. Users take notes, make annotations and sketches, or interact with printed user interfaces, e.g. by tapping with a pen on printed buttons. There is a wide range of technologies for capturing pen input. Technology for capturing pen input on real paper should offer high tracking performance while restricting the natural interaction as little as possible. A first class of approaches tracks the relative position of a pen with respect to a separate tracking device, e.g. a camera. A second class of capturing approaches is able to directly capture the absolute position on a page.

All approaches have in common to generate not only a two-dimensional representation of the pen traces (as does a photo or a scan of the document). In addition, they record temporal information of how the traces are made over time. Some technologies also track the force with which the pen tip is pressed onto the paper sheet. At regular intervals, e.g. 50 times per second, a so-called sample is captured. This sample contains the current position of the pen, the current time and optionally the pen tip force. A set of samples is called digital ink data. By interpolating curves through the sample coordinates, the pen traces can be visualized.

Camera-based Capturing

Above we have discussed how camera-based capturing can be used for detecting touch input. Similar approaches have been presented for tracking pen input. The DigitalDesk [167] analyzes images of the desk that an overhead camera is continuously capturing. It detects not the pen itself, but the traces that are made with a pen. While this allows the system to record traces, it is not possible to detect interactions that leave no visible ink traces, such as tapping with the pen.

Other approaches aim at tracking the pen itself, not only the visible traces the user has made with it. Typically one or several visual markers are attached to the pen. This allows the system to robustly identify the pen in the camera image. These markers can be active markers that emit light signals [70] or passive markers that reflect light of a given wavelength [40].

Ultrasonic Pens

A second technique relies on ultrasonic triangulation. The digital pen continuously emits an ultrasonic signal which is not audible by humans. An additional separate tracking device is attached to the paper sheet(s). It has two or more reference points that capture the ultrasonic signal. By calculating the time difference with which the signal reaches the reference points, the position of the pen with respect to these reference points can be calculated.

The temporal and spatial resolution of this technology is high enough for capturing handwriting. The tracking does not depend on the material the pen is used upon and it scales to large surfaces. So it can be used on arbitrary flat surfaces, such as tables, augmented walls or whiteboards. However, the position of the pen is tracked in relation to the external device and not in relation to the paper sheet. Moreover, this approach is not able to detect which sheet the user is writing on. This makes this approach hard to use in settings where users do not write on one single page but deal with many pages. Ultrasonic tracking is utilized in commercial solutions that target private end-users, for example the Pegasus Tablet NoteTaker[8]. It has a resolution of 100 dpi.

Digitizing Tablets

Digitizing tablets are devices that capture pen input for computer applications. The most common technology is patented by Wacom. A Wacom device has a flat surface and generates a magnetic field. Using induction, the position of a specific stylus can be detected on this surface. Figure 2.5 shows a Wacom Intuos4 graphics tablet[9]. In principle, digitizing tablets do not aim at supporting the use of a pen on paper. Instead, the tablet is used to directly interact with the digital system, e.g. for drawing in a graphics application or for positioning the mouse pointer. However, the induction principle still works if one or several sheets of paper are positioned between the graphics tablet and the stylus. For this reason, they can be used to track pen input on real paper. This approach enables very high resolutions (about 1000 to 5000 dpi). However, when used for capturing pen input on paper, it has the same drawbacks as ultrasonic tracking. The user must manually calibrate the paper sheet and must indicate page changes. Moreover, the interaction is restricted to the small surface of the tablet. Therefore this approach is typically used only in research prototypes (e.g. [88, 27, 150]).

Inductive Pens

The following techniques do not detect the position of the pen with respect to an external reading device, but they encode positional information directly on paper. Thereby local page coordinates can be detected without requiring the system to have knowledge about the absolute position of the sheet of paper. It is the digital pen that decodes its position, which makes external devices and calibration obsolete. As the position which is encoded on the physical sheets can also contain a page identifier, the pen is able to detect on which page it is used. Hence, it is not necessary to manually indicate on which page the pen is currently used. Users can therefore work very naturally with multiple sheets of paper.

Within the frame of the European Paper++ project, researchers have developed an approach that leverages conductive ink [134, p. 26]. Positional information is

---

[8] http://www.pegatech.com

[9] http://www.wacom.com

2.1 Technologies

Fig. 2.5 Wacom Intuos 4 digitizing tablet (photo copyright Wacom)

encoded directly on paper by printing a grid of linear barcodes with conductive ink on paper documents. This ink is barely visible to the human eye. A digital pen measures the inductivity and decodes from the barcodes its two-dimensional position. The resolution of this technique is not high enough to capture handwritings but it allows for distinguishing larger hot-spot areas on a document.

Anoto Technology

Anoto digital pen and paper [4] is the currently most mature solution for capturing pen input on paper. An Anoto digital pen behaves like an ordinary ballpoint pen and leaves visible ink traces. Moreover, it has a built-in infrared camera and a processing unit to detect a specific pattern on the print products. By analyzing this pattern, the pen can decode its position and electronically capture all pen traces made. This is done at a frame rate of approximately 75 Hz and with a spatial resolution of about 850 dpi. In addition to the position in the two-dimensional coordinate space, current pens register pen tip force and timestamps. The technology further allows for detecting the rotary and tilt angles of the pen although this is not supported by the firmware of current pens. Anoto pens include a battery and can therefore be used in mobile conditions. An Anoto pen is depicted in Fig. 1.2 on p. 12.

Depending on the capabilities of the pen, data is temporally buffered on the pen until it is synchronized with a computer via USB or Bluetooth. Alternatively the data is streamed in real-time to a nearby computing device using a Bluetooth connection.

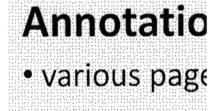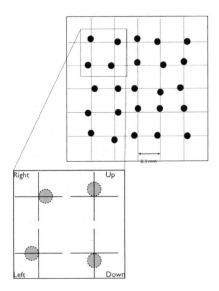

Fig. 2.6 The Anoto pattern. Left: Enlarged view of a document that contains the pattern. Right: Highly enlarged schematic view of a 6x6 grid of points of the Anoto pattern (illustration copyright Anoto)

Unfortunately, it is not end-users who can decide in a given situation whether the pen should temporally store the data or stream to a nearby device. Instead the designer of the application has to select beforehand which of the modes to use. This is necessary because two different types of Anoto pattern, which have to be licensed separately, enable either batch processing or streaming, but not both of them.[10]

Anoto Pattern The positional information is encoded directly on the paper sheets with a patented two-dimensional dot pattern (Fig. 2.6). The pattern is only slightly visible to the human eye, making the print product appear slightly gray. Each dot has a position on an imaginary grid which overlays the page. By slightly displacing each dot in one of the four directions, a single dot encodes 2 bit of data. Each 6x6 matrix of dots encodes a unique position in the Anoto coordinate space. This gives not only the position on a page, but also allows for distinguishing between different pages.

The Anoto pattern can be printed on paper products with various printing techniques, including offset, laser and inkjet printing. Printing is unproblematic if the dot pattern is to be printed on empty sheets of paper that do not contain any other printed contents. In this case, most desktop laser printers can be used, and also many inkjet printers yield acceptable results. Things get more complicated if the page contains printed contents on which the Anoto pattern is overlaid. The problem here is

---

[10] To our knowledge the only exception is the Nokia SU-1B pen, now discontinued, for which a firmware patch exists that enables switching between batch processing and streaming on non-streaming pattern. This pen is used in many research prototypes.

that the camera might not see the pattern dots at positions where other content is printed. To cope with this problem, the Anoto technology relies on a smart solution that leverages an optical characteristic of toner and ink. Some toners absorb infrared light and thus appear in the image of the infrared camera in black. These toners are used for printing the dot pattern. Other toners do not absorb infrared light. While these toners are visible to the human eye, they are not visible in the camera image. These toners can be used for printing contents other than the dot pattern. These contents do not interfere with the dot pattern in the camera image. The printer requires toner (or ink) that has these characteristics. It happens that the black toner (K) used in many laserjet printers typically absorbs infrared ink, while cyan (C), magenta (M) and yellow (Y) toners do not. This allows printing the dot pattern with K toner, while other contents are printed by using only C, M and Y colors. The Anoto pattern and other contents can be printed in two separate steps or in one single step. Anoto has tested printers of diverse manufacturers and has published a list of Anoto-certified printers. If the pattern is to be applied to larger paper products, it can be printed with some selected inkjet plotter models.[11]

Pen Models Anoto pens are produced by several manufacturers. The Anoto/Maxell DP-201 and the Logitech/Destiny io2 support USB batch processing and Bluetooth streaming. The Anoto ADP-301 supports only Bluetooth streaming, but features a lower latency than the DP-201. This makes it the product of choice for applications with hard real-time constraints on input. Finally, the Anoto ADP-501 features a thicker pen tip instead of the thin ballpoint tip. This provides for use on flipcharts and whiteboards. Several older pen models are discontinued, most notably Logitech io and io2, Nokia SU-1B and Nokia SU-27W.

These standard models suffer of several shortcomings that will be discussed in the following: pen feedback, erasing of contents and use as a stylus on displays.

Pen Feedback The standard models do only capture and digitize pen traces to transfer them to a computer. They do not interpret them and therefore cannot provide system-specific feedback, even though all pens feature LEDs and some of them can vibrate. Moreover, the Bluetooth connection does not offer any back-channel to control the pen, so applications cannot generate feedback given by the pen. Liao et al. [76] extended one of these pen models with auditory, tactile and visual feedback and explored with that prototype several pen-based feedback mechanisms. Recently, several new pen models have been commercialized that include additional processing capabilities and output devices. These pens are able to interpret pen traces and react to user input. The Fly Fusion Pentop Computer[12] has a built-in speaker. It reads out the translations of handwritten words, can be used as a calculator and plays MP3 files. The more recent Fly Tag Reading System aims at helping kids learn to read. It features a digital pen that reads out the passages of a book the user points to. The Livescribe Echo Smartpen[13] (Fig. 2.7) has a built-in microphone, speaker and an

---

[11] For instance the plotters of the Canon imagePROGRAF series (iPF8000, iPF8300, iPF9000, iPF9100) provide quite good results on diverse types of paper and foils.

[12] http://www.flyworld.com

[13] http://www.livescribe.com

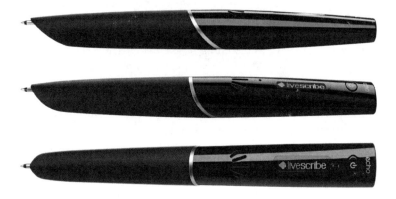

Fig. 2.7 Livescribe Echo SmartPen (photo copyright Livescribe)

OLED display. In addition to capturing notes, it can be used to record and playback audio and to perform calculations. An SDK provides for developing further applications (so-called penlets) which are executed directly on the pen. While these novel pens enable to develop more interactive interfaces including direct system feedback on the pen, they do not provide a wireless Bluetooth connection.

Erasing Pens A further shortcoming of current Anoto pens is that they do not allow users to erase physical pen traces. Olberding and Steimle [111] have demonstrated an easy and inexpensive method to add erasing capabilities to Anoto pens. Anoto plans to commercially deploy an eraser pen (ADE-501) in the near future that can be used to erase contents on whiteboards, but not on paper documents.[14]

Pen Input on Displays All these solutions aim at supporting pen input on paper. However, Anoto pens can also be used as a stylus for providing input on screens. Brandl et al. [10] presented an approach that enables using Anoto pens on rear-projection multi-touch screens. This type of screen is typically used in interactive tabletop and wall displays. Their approach relies on printing the Anoto pattern on a translucent foil. Like on paper, the pen decodes its position from the printed pattern. This position can be converted to screen coordinates. More recently, Hofer and Kunz [38] introduced a method that allows using Anoto pens also on LCD displays. Anoto is announcing their own solution for input on displays to be released soon. Theses approaches enable to produce very large pen-enabled displays that have a high input resolution at relatively low cost. Moreover, in contrast to competing technologies for pen input on displays, multiple pens can be simultaneously used.

---

[14] http://www.anoto.com/up-and-coming-products.aspx

Data-embedding Pens

A recent direction of research examines data-embedding pens [81]. In addition to leaving visible ink traces, a data-embedding pen visually encodes digital data onto the paper document. Liwicki et al. present a prototype device that features, in addition to an ordinary ink refill, a tiny ink-jet head. This ink-jet head can print a sequence of dots in a bright color close to the handwritten traces. This dotted line encodes a sequence of binary data, which can be subsequently decoded by using optical image processing. A data-embedding pen can be used for instance for encoding metadata of the handwriting, such as the time of writing, the ID of the author or the geo-coordinates of the current location. Alternatively, the pen can encode the temporal sequence in which the pen traces have been written. Such information about the writing process can improve recognition of scanned handwriting.

2.1.5 Digital Output on Paper

We have discussed a variety of approaches that allow systems to capture paper contents and input made on paper. To conclude the section on technologies, we will briefly discuss how systems can provide *output* directly on paper. While a straightforward way consists of (re-)printing a document, this is not very interactive, since each update requires a new printout. We will have a look at technologies that enable more dynamic output on paper.

Projection

The projection approach uses paper as a passive display onto which digital contents are projected. The position and orientation of the paper surface is tracked in real-time. The system then projects the display contents onto this surface. This enables not only to simulate screens, but also to overlay physical objects with additional digital information. If the performance of the tracking system is sufficently high and the projected image is correctly adjusted to fit the physical surface, the paper surface can be moved around freely and behaves similar as if it was an active display. However, even in state-of-the-art solutions, the projected image is slightly lagging behind the paper surface if the surface is moved quickly.

This approach is relatively straightforward if paper is used only on a flat surface. With the influential Augmented Surfaces system [123], Rekimoto and Saitoh demonstrated that a single camera and a standard projector are sufficient for tracking paper on a 2d surface. If the user is allowed to freely move and rotate paper surfaces in three dimensions, things get more complicated. In this case, the surface is not necessarily perpendicular to the orientation of the projector. So the projected image has to be distorted such that the projected image appears undistorted to an observer. Early examples are Dynamic Shader Lamps [7] and PaperWindows [40]. Most solutions use commercially available optical motion capture systems, such as Vicon or NaturalPoint OptiTrack. A set of several high-speed cameras detects the 3D positions of retro-reflective makers that are attached to the paper surface. This allows the system to identify the position, orientation, and possibly deformation of the surface. Recently Lee et al. [70] have demonstrated that even a low-cost tracking solution, using the PixArt camera within the Nintendo Wii Remote, can yield acceptable results for 3D tracking. The Microsoft Surface team presented a different approach for projecting contents onto passive displays [47]. Instead of top-projection, they use a sophisticated setup for projecting contents onto semi-transparent paper surfaces from the backside. For this purpose, a projector is integrated into an interactive tabletop display. The tabletop surface consists of an electronically switchable diffuser. This can alternate with very high frequency between a diffuse and a transparent state. During the diffuse time slots, imagery is projected onto the tabletop surface. During the transparent time slots, imagery is projected through the tabletop surface onto the paper displays.

Electronically Augmented Paper

Output on paper can be realized by attaching or embedding electronic components into paper. This enables a greater diversity of interactions than with passive paper, but electronically augmented paper documents are more complicated and more expensive to produce.

An example is *Pulp-based computing* [17]. The authors embed diverse components directly into paper during the papermaking process. These components comprise microphones, bend sensors, LEDs, speakers and vibrating motors. *Voodoo-Sketch* [9] is an ad-hoc physical interface toolkit. It allows plugging various electronic controls, such as push buttons, switches and sliders, onto specific paper palettes, which have embedded conductive layers. The *Computational Sketchbook* [11] follows a similar approach. The user can create interactive paintings by attaching electronic components onto ordinary paper. These components comprise speakers, motors, switches, LEDs and batteries. The user establishes the conductive connections by painting them with conductive ink. The electric circuit thus becomes a visible part of the artwork.

Electronic Paper

Upcoming novel display technologies have a large potential to significantly alter the way we are interacting with paper and with displays. Rapid advances in OLED and Electronic Paper technologies allow companies to develop displays that have similar characteristics as paper. Current commercialized displays, e.g. displays of recent e-book readers, are still rigid and much thicker than traditional paper. However, various research prototypes show promising advances. Japanese researchers have developed an electronic paper display which is as thin as 1 mm; Sony has developed

2.1 Technologies 35

an OLED TV of only 2 mm thickness [54]. Moreover, Sony [141] and Polymer Vision [120] have presented displays that can be rolled. Recent research has even used paper as the substrate of an electronic paper display. This is much thinner than traditional electronic paper and more flexible and soft [54]. Such displays will allow designers to develop applications that leverage the affordances of paper while offering full digital feedback comparable to a screen.

2.1.6 Pen-and-Paper Toolkits

As Anoto pens are the currently most mature and most widely used solution for realizing Pen-and-Paper Applications, we present some toolkits that ease developing applications featuring the Anoto technology. All of these toolkits offer support for establishing the low-level connection between the pen and a computing device and for accessing the pen data. Some toolkits additionally provide higher-level support for interpreting pen data.

Anoto SDK

Anoto offers commercial SDKs for developing Pen-and-Paper Applications with Anoto pens. Only little information about the SDKs is made publicly available. For guiding application developers, we outline core concepts of the Anoto SDKs.

Developers have to license some Anoto pattern space that can be used within the application. The entire Anoto pattern space is large enough to cover a surface equivalent to that of Europe and Asia combined. A license consists of a subregion of that space. For instance, a "book" license is equivalent to 256 letter-size or A4-size pages that each can be uniquely identified.

The Anoto pen delivers raw coordinates which describe a unique location in the entire pattern space. In typical applications, these huge coordinates are quite cumbersome to handle. One is rather interested in coordinates that are relative to a given document or a given page. The Anoto SDKs provide such higher-level abstractions of the raw pen coordinates. For instance, it is easily possible to retrieve all traces that were made on a specific page. The SDKs further allow application developers to define interactive regions on document pages. For instance, a developer might want to use a paper form in the application. This form might have some preprinted areas into which the user can indicate some information, e.g. for the user's name and address. In addition it might contain some boxes that the user can optionally check, e.g. indicating if she wants to receive further information. All these areas can be defined as interactive regions. On a technical level these are rectangular areas with an associated identifier.

To define such interactive regions on paper products, Anoto offers the Form Design Toolkit, a plug-in for Adobe Acrobat. This allows for defining interactive regions and adding the Anoto pattern to any PDF document. However, the Form

Design Toolkit requires manually executing several commands within Adobe Acrobat. Manually generating these documents is not problematic if an application uses a limited number of paper documents that are identical for each user. However if users need personalized documents, possibly even creating and printing them during runtime, the manual approach is inadequate. For such cases, Anoto offers the Paper SDK. This SDK enables applications to add the Anoto pattern to documents during runtime.

For developing applications, Anoto offers an SDK for PC applications (including Java and for native Windows applications) and an SDK for network applications. Both SDKs support only non-streaming applications, offering classes and methods for accessing and parsing pen data that has been transferred via batch synchronization. The SDKs allow the developer to directly retrieve pen traces that were made on a specific interactive region. Furthermore, the SDKs offer basic additional functionality, such as for rendering and for saving pen data.

Access to the streaming functionality of the ADP-301 is supported by a further SDK, the Streaming Pen Connectivity Driver and its accompanying API. This provides basic methods for accessing the pen data arriving through the Bluetooth connection. While it gives access to the linear data stream, to the best of our knowledge it does not provide the abstractions offered by the non-streaming SDKs described above.

PaperToolkit

The PaperToolkit [178] is an open source toolkit developed by Ron Yeh at Stanford University. It builds on top of the Anoto SDK. PaperToolkit aims at an interactive use of digital pens in applications and provides more support and higher-level abstractions of digital pen data than the Anoto SDK. It is based on an event-based architecture, similar to frameworks for Graphical User Interfaces toolkits like Java Swing or Windows Forms. The application developer can compose a paper interface by composing predefined interface widgets, such as drawing areas, buttons and check boxes. The toolkit renders these widgets on top of the Anoto pattern to a PDF document and allows for printing it on paper.

For applications the framework provides event handlers that react to specific types of pen input. Event handlers can be added to paper interface widgets and trigger events on the software side. For instance, the application can register event handlers that are invoked when any pen activity has occurred within a specific widget, when a button widget has been tapped on or when a check box has been selected.

PaperToolkit unifies real-time and batched event handling. It is also possible to use several pens simultaneously in the same application. The toolkit contains plenty of further components that ease developing interactive Pen-and-Paper Applications. This includes a component that allows developers to define a fully functional paper interface simply by sketching it on paper, diverse components for rendering digital pen data as well as a tool for simulating pen data for debugging purposes.

2.1 Technologies   37

The toolkit is developed in Java. We have used it in our own projects and have experienced that it is very stable. However, it is not under active development any more. While older pen models (Logitech io2, Maxell DP-201, Nokia SU-1B and SU-27W) are supported, more recent pen models cannot be used with the toolkit unless extra code is written to access these pens.

iPaper

The iPaper framework [134] was developed by Beat Signer and his colleagues at ETH Zurich. It supports the development of interactive paper products by combining an authoring perspective with a service approach. The underlying concept is to decouple visual design, interaction design and development of the services. The graphic designer authors the document. A service developer implements a service that can be used with an interactive paper document. Finally, the interaction designer ties both the document and the services together by defining links between regions of the paper document and services.

iPaper is based on iServer, a generic framework that allows the definition of links between arbitary physical or digital information entities. For instance, links can be defined between any combination of images, videos, digital documents, paper sheets, RFID tags and many other types of resources. Each link can have multiple source and multiple target resources. The iPaper plug-in for iServer provides more specific support for Pen-and-Paper Interfaces. iPaper links are established between an active region on paper and an active component, which is the service associated with this region. In contrast to the Anoto SDK, active regions do not have to be rectangular. Other shapes, such as circles and polygons are supported. When the pen is used on an active region, the framework automatically executes the code of the active component that is linked to this region. While a number of predefined active components are available, the application developer can easily add own active components.

iPaper has a distributed architecture. For instance, an active component can be deployed on a server. Moreover, it includes a powerful printing component. This component allows printing documents that contain the Anoto pattern directly from within the end application, similar to the Anoto Paper SDK. It supports most of the current pen models (including Anoto ADP 201, Logitech io2), but not the most recent Anoto ADP-301.

Letras

Letras [36] is an open source toolkit that provides support for mobile Pen-and-Paper Interfaces in ubiquitous computing settings. Letras enables to use the same pen in a variety of different contexts, in combination with various computing devices. Perhaps most interesting is that it allows users to couple Anoto pens with Android phones and tablets. This provides for using interactive Pen-and-Paper Applications nearly everywhere. In this case, pen data is streamed to a personal mobile device that acts as an information hub and decides how to distribute pen data further to other computing devices over a network connection. Letras also supports different configurations; for instance the pen can be coupled with a PC.

Letras has a modular architecture. It relies on a distributed processing pipeline for pen data. Different processing stages are decoupled using generic interfaces. The first processing stage, the driver stage, connects to one or several Anoto pens. This includes simultaneous connection to different pen models. A next processing stage provides for abstracting from raw pen coordinates to higher-level interactive areas. This is similar to the interactive areas of the Anoto SDK, but with a distributed lookup of area definitions across multiple issuing organizations, allowing to distribute interactive paper material such as leaflets. Further processing stages support semantic processing, e.g. handwriting recognition, and application-level processing, e.g. rendering the ink traces. Application developers can develop additional processing stages and flexibly deploy processing stages to different computers and devices. While the first processing stage is typically deployed on a computing device that is used by end-users (e.g. the mobile phone, the tablet or the PC), the application developer can decide where to deploy other stages. For instance, handwriting recognition might be performed on a server. For rendering, the system might detect which computing devices in the room feature a large screen and dynamically deploy the rendering stage on one of these devices. The individual stages are interconnected with the MundoCore middleware [3]. This supports the dynamic configuration of the computing environment. A drawback of Letras is that it currently does not provide a print toolkit. For adding the Anoto pattern to documents, the developer must use one of the Anoto SDKs.

Letras is developed in Java and contains some native components for Mac OS X, Windows and Linux. It provides native support for streaming pen data to Android devices. It currently supports Anoto ADP-301, Anoto DP-201, Logitech/Destiny io2 and Nokia SU-1B.

Livescribe SDK

Livescribe offers an SDK for their Livescribe pens.[15] It can be downloaded and used free of charge. It follows a different approach than the frameworks presented above. In contrast to standard Anoto pens which do only capture pen data and transfer it to a computer, a Livescribe pen has a processing unit that can execute custom code. Livescribe applications consist of a paper product and a penlet. A penlet is a piece of software that is executed directly on the pen. This enables to develop applications that react on user input in real-time without requiring a streaming connection to a second computer device which handles the interpretation. Furthermore, a specific Desktop SDK offers support for developing desktop applications that access pen data (this is similar to the standard way of Anoto-based applications).

---

[15] http://www.livescribe.com

The Livescribe API is based on the Java Platform Micro Edition (Java ME). It comes with Windows and Mac versions of an Eclipse-based IDE and a software for emulating the Livescribe pen.

Livescribe has a management of paper regions that is different from the other toolkits. It distinguishes between fixed print and open paper. Fixed print areas are similar to the interactive regions of the other toolkits. They allow developers to define specific areas for widgets that control the application. In contrast, open regions are blank regions of paper that are not permanently associated with one specific application. Instead each application can claim an open region when the user is writing on it. This enables the flexible use of one single paper notebooks for many different applications.

2.2 Pen-and-Paper Interfaces

A wealth of applications and interaction techniques of Pen-and-Paper Interfaces have been presented in prior work. In this section, we review the large and ever-growing body of research that has been established during the past two decades.

Early, highly influential research was conducted at XeroxPARC and EuroPARC in the early 1990s. A number of seminal systems demonstrated how paper can be closely integrated with digital media on one single interactive tabletop surface. These systems already showed three central functions of augmented paper: 1) paper is used as a token for accessing and controlling a digital resource, 2) paper documents contain embedded hyperlinks to additional digital resources, and 3) the contents of a paper document are automatically synchronized with a digital version of the document. These systems laid the foundation for a wide range of paper-augmented tabletops and walls and also for a number of approaches that allow users to manage digital media on their PC by using paper tokens.

While in this first generation of works, paper was mostly restricted to be used solely on the interactive tabletop or on the user's desk, a subsequent generation focused on paper as a mobile medium. Augmented paper notebooks, form-filling applications and applications for annotating printed documents aimed at retaining the freedom to use paper at many different places. The advent of digital pens which can be used in mobile settings without a complicated technical setup certainly promoted this evolution. In contrast to the early desk systems, most of these mobile systems provide visual output not directly on paper, but on a separate computer screen. However, we will also review approaches that realize visual output directly within paper documents by using mobile projectors or tiny overlaid displays. Support for paper-based collaboration is an issue of more recent interest. This comprises not only co-located collaboration, but in particular how asynchronous remote sharing of paper-based contents can integrate paper with social networks and the Web 2.0.

Our review is structured following which paper media are augmented by digital functionality and following the main functions of augmented paper. Augmented paper cards and post-its clearly show how paper can be used as a token for accessing and controlling digital resources. Augmented books demonstrate paper-digital hyperlinking. Augmented paper notebooks and augmented printed documents present a variety of approaches to synchronize paper-based contents with a digital version. Finally, augmented tables, flipcharts and walls integrate all of these functionalities and provide a very seamless integration of physical and digital media on large surfaces. We will conclude this chapter with a discussion of future directions of research.

2.2.1 Augmented Paper Cards and Post-Its

Augmented paper cards and augmented post-its demonstrate the first main function of augmented paper: using paper as a physical token for accessing and managing digital resources. Each digital resource to access is represented by a physical object. This object has no functionality other than representing the digital resource; it does only contain a description of the resource, but not its actual contents. By manipulating this object (e.g. by holding it in front of a barcode reader or near to an RFID reader or by pressing a push button on that object), the associated digital resource is accessed. This provides an easy and intuitive way for selecting and opening documents or applications. Moreover, the physical tokens can be flexibly structured by arranging them in space and can be physically shared with co-workers.

An example is *WebStickers* [82] which uses post-it stickers that contain a barcode. The user can associate a post-it note with a single Web page or with a collection of Web pages. By holding the post-it under a barcode reader, the associated resource is displayed on a computer screen. *Palette* [101] follows a similar approach and uses barcode-enhanced paper cards for accessing individual slides of slide presentations.

PaperButtons [114] overcomes the limitations of barcodes that must be placed below a barcode reader or in front of a camera phone. It extends the Palette system by replacing the barcode on a paper card with a push button. A unique ID is transmitted to the system over a wireless connection when this button is pressed.

The previous systems allow easy access to the digital resource, but do not support the opposite direction: easily finding the physical token that belongs to a digital resource. To support this opposite direction, *Quickies* [95] augments physical post-it stickers with an RFID tag that is applied to the reverse side. The user can write a handwritten label on a sticker using a digital pen. The label is automatically digitized, which allows for searching through all stickers in a desktop application. the corresponding physical stickers can be found using an RFID reader. *Move-It* [121] further improved this approach. The system can electronically actuate individual Post-it stickers. This supports not only easy finding of a specific physical sticker, but also allows the system to trigger physical notifications. For instance, a post-it which contains a reminder for a meeting can be actuated shortly before the scheduled time of that meeting.

2.2.2 Augmented Books

Augmented books demonstrate the second main function of augmented paper: enriching printed contents by embedded hyperlinks which point to additional digital resources. Thereby classical reading of a physical book can seamlessly evolve into browsing a physical-digital hypertext. Augmented books are conceptually similar to paper tokens in that they link from paper to digital resources. However in contrast to a paper token, which merely acts as a physical representation of a digital resource, an augmented books has its own value, independently from the digital resources. Here, hyperlinks do not reference between two instantiations of the *same* resource, but between *different* resources.

We can classify these systems following the technology which is used to select hyperlinks on paper (and the resulting interactions), the type of the digitally linked media, and the device on which this digital media is made available. The underlying interaction metaphor is similar in all thee systems: selecting a link hot-spot with a deictic gesture for accessing the associated digital resource. Depending on the technology, the deictic gesture is realized by tapping with a stylus or the finger, clicking with the mouse, or scanning a marker with a camera or a barcode reader.

ActiveBook [137] consists of a paper book and a specific point-and-click selection device that contains a barcode reader and a mouse. Pages of the book contain active areas that serve as link hot-spots to Web resources. These areas can be of various shapes and sizes, allowing for very flexible link anchors. A hot-spot is selected by moving the selection device over it. The system does not automatically recognize which page of the book is open. For this purpose, each page contains a linear barcode that the user has to scan before selecting a link hot-spot.

The *Listen Reader* [6] is an augmented physical book with embedded hyperlinks to audio files. It conceptually improves over ActiveBook in two aspects. First, it does not require any interaction device but allows the user to select links by simply tapping on a link hot-spot. Second, the current page of the book is automatically detected using RFID technology. A passive RFID tag is embedded in each page of the book. An RFID reader in the back cover of the book detects which pages form the right hand side of the opened book and infers the current page. For detecting touch and hover interactions, the Listen Reader relies on inductive sensing.

Books with Voices [62] is an augmented paper book that provides links to video resources. Each link is represented by a linear barcode that is printed on the page margin besides an interlinked paragraph of text. When the user wants to follow this hyperlink, she scans the barcode with a PDA device that features a barcode reader (Fig. 2.8). The PDA identifies the target resource and plays back the video. While the interaction is less direct than with ListenReader and the large number of barcodes potentially interferes with the visual design, Books with Voices has the advantage to be easily deployed in real settings using standard hardware. Moreover, digital contents can be displayed in-situ using on the PDA. Today, a similar concept is widely deployed: QR codes [46] are printed on paper documents that encode the URL of a digital resource. The user can easily display the resource on a camera phone by reading the barcode with the camera.

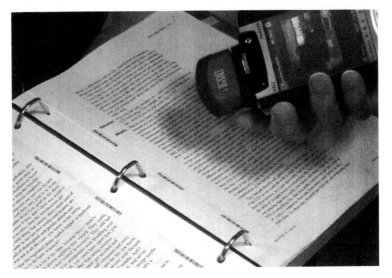

Fig. 2.8 Books with Voices (photo courtesy of Scott Klemmer)

The advent of Anoto pens enabled realizing augmented books that do not require visual markers nor an extensive hardware setup. *Print-n-Link* [108] supports easy retrieval of references in scientific articles by pointing with an Anoto pen on a printed reference. The *LeapFrog Tag Reading System*[16] (Fig. 2.9) is a commercial augmented book. The system turns a storybook interactive with the goal to help children learn to read. An Anoto pen, which is redesigned for kids, can be moved over the pages of the book. Audio information related to the selected content is then played back on the pen. For instance, words are read out. Moreover, the pen automatically logs information about how it is used. This information can be accessed by parents or teachers to examine the kid's learning progress.

These systems offer appropriate support as long as the focus is on reading a document and it is sufficient to have access to some pre-programmed digital resources. However, if the user has a more active role and wants to create user-defined hyperlinks, pre-programmed links are too restricted. We will reviews approaches that allow users to define own hyperlinks on printed documents in Section 2.2.4.

2.2.3 Augmented Paper Notebooks

The third main function of augmented paper consists of synchronizing a paper-based with a digital version of the same resource. Augmented paper notebooks are an important class of interfaces that apply this principle. Contents of a traditional, paper-based notebook are automatically digitized and made available on a computer. This

---

[16] http:/www.leapfrog.com

2.2 Pen-and-Paper Interfaces

Fig. 2.9 LeapFrog Tag Reading System (photo copyright Anoto)

retains many of the advantages of a paper notebook: the user can write and sketch freely with a pen, the notebook is mobile and it can be used in a variety of settings, including casual notetaking in cases when using a computer would be inappropriate, think for instance of a therapy session. Moreover, the paper notebook provides rich physical cues for navigating through the contents.

All pen traces are automatically captured in an electronic form and transferred to a computing device, either continuously in real-time or at specific points in time when the user synchronizes the pen with a computer. A software viewer allows the user to browse through a digital facsimile of the paper notebook. Subsequent updates made on the paper notebook are automatically integrated into the existing digital version. The digital version has the advantage to be searchable, e.g. by time. It is easily accessible even if the physical notebook is not available, for instance for the purpose of archiving. In addition, some augmented notebook systems automatically perform handwriting recognition of the pen traces. The recognized text is typically not displayed, but used in the background to enable full text search within the notebook.

Nowadays, this basic functionality of an augmented paper notebook is provided by many commercial solutions for end-users. Examples include the Livescribe Echo smart pen software[17] and the Oxford Easybook notebooks[18]. While these solutions

---

[17] http://www.livescribe.com

[18] http://www.oxfordeasybook.com

come with their own viewer software, the software Adapx Capturx for OneNote[19] automatically integrates handwritten notes and sketches into Microsoft OneNote. In addition, some augmented notebooks (e.g. Oxford EasyBook software) allow the user to select commands to be performed once the notes are transferred to the computer. For instance, a note can be automatically sent to a specific person by electronic mail or it can be added as a new task to Microsoft Outlook's task list. These commands are invoked on a note by writing a specific sign (e.g. an encircled letter) next to the note.

Integrating Additional Media

More advanced research prototypes demonstrate how the basic principle of synchronization can be combined with more advanced functionality, such as hyperlinks to additional digital resources. Influential in this respect was a series of research prototypes that aim at supporting biologists. Biologists make frequent use of notebooks for jotting down information when they are in the field, e.g. while observing species or collecting specimen, and also use their notebooks when they are in the lab. Many of these notes are closely related to other resources, such as articles, photos or physical specimen. While these typically remain separate from traditional paper notebooks, augmented notebooks allow biologists to integrate these different sources of information closely with their notebook. As a matter of course, such augmented notebooks do not only support the effective work of biologists, but equally apply to a wide range of professional occupations. For instance, also designers typically make heavy use of notebooks.

The *A-Book* [88] is an augmented lab notebook for biologists. It comprises the basic functionality for accessing paper-based notes via a computer interface that was introduced above. In addition, it contributed a set of important concepts for augmented notebooks. First, the A-Book introduced digital means for integrating the paper notebook with external resources. The user can create hyperlinks to Web pages and can paste physical objects, such as printouts and photos, into the paper notebook, provided that these are "known" to the computer. Second, the A-Book offers functionality for structuring the notebook. The user can easily define a digital table of contents by adding graphical snapshots of passages of the notebook which are listed in chronological order. Moreover, it is easy to connect different passages within the notebook by creating hyperlinks.

A-Book's probably most influential contribution is the Interaction Lens. The Interaction Lens makes digital information (such as hyperlinks) available in-situ, directly within the notebook, instead of on a separate computer screen. The Interaction Lens consists of a PDA, which can be placed onto the paper notebook, moved and rotated. The Interaction Lens appears to be transparent, visualizing on its display the physical contents of the notebook that it is occluding (Fig. 2.10). The position and orientation of the PDA is automatically tracked. This enables the system to maintain

---

[19] http://www.adapx.com

2.2 Pen-and-Paper Interfaces

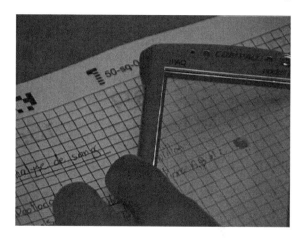

Fig. 2.10 Interaction Lens of the A-Book for accessing digital content directly within the physical notebook (photo courtesy of Wendy Mackay)

the illusion of a transparent PDA for any orientation. The Interaction Lens acts as a window between the physical and the digital documents. It overlays the physical contents with additional digital information, such as hyperlinks. It is not only used for visualizing contents, but also for interacting with the notebook, such as for creating hyperlinks or for adding passages to the table of contents. Hence, the Interactive Lens seamlessly integrates physical and digital information, but only on the small screen of the PDA which must be moved over the page to successively view the digital contents of the entire page. While the concept of the Interaction Lens *per se* is mobile, the A-Book prototype is not, since it relies on a stationary technical setup for capturing pen traces and the position of the PDA, using a digitizing tablet.

ButterflyNet [176] is an augmented notebook for field biologists. It improved upon the A-Book by contributing a set of interaction techniques that allow users to easily link from the notebook to digital resources without requiring a computer. For instance, photos taken with a digital photo camera are automatically added to the digital version of the notebook and aligned with the handwritten contents using their timestamps. Optionally the user can also perform a specific pen gesture to integrate a photo at a specific position within the digital version of the notebook. Moreover, it is possible to easily create links to physical objects, such as specimen collected in the field. The user has first to place the object into an envelope that a barcode is printed on. By taking a photograph of the barcode, the link is established. An important advantage here is that all activities can be done with the paper notebook – no manual post-processing of the digital version is necessary. The ButterflyNet software viewer then provides access to a digital version of the notebook that includes the photos (Fig. 2.11 left). *Memento* [168] is a hybrid physical and digital scrapbook that uses a similar concept for embedding pictures and videos.

Prism [149] is an augmented paper notebook that also targets biologists. The focus is here on the feature-rich Web-based digital version of the notebook into which handwritten notes are automatically integrated. In addition to a digital copy of the handwritten notes, the Web-based version can contain typewritten notes as well as hyperlinks to Web pages, e-mails and local documents or snapshot images

 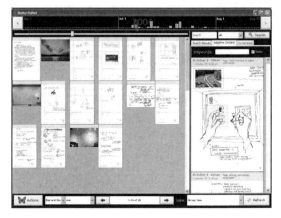

Fig. 2.11 The ButterflyNet/IDeas notebook [176, 69]. Left: Paper notebook. Right: Shared digital notebook including selected pages of multiple users (photo courtesy of Scott Klemmer)

of them. In contrast to ButterflyNet, additional media cannot be integrated using the paper notebook but only in the Web-based interface. The authors conducted one of the most extensive studies that examine how Pen-and-Paper User Interfaces are used. Prism was evaluated in a long-term study over a period of nine months with 5 participants. Amongst others, the authors found that the paper notebook was preferred to be used as a master notebook. This is an organized account of what the users did and planned to do – a central information hub that systematizes and integrates information from various sources.

Another stand of research examines how handwritten notes can be coupled with temporal data, such as audio or video recordings. The *Audio Notebook* [148] closely couples handwritten notes with an audio recorder. It comprises a digital audio recorder and a paper notebook that is placed on top of a digitizing tablet. Handwritten notes are automatically captured in a digital form and used to index the temporal audio stream. If notes are taken while audio is recorded, the notes are automatically linked to the temporal position within the audio stream that corresponds to their creation time. By tapping with the pen on a note, the audio recording is played back at that specific position in time. This is a lightweight, yet powerful indexing mechanism that lays the temporal audio stream out in space. This concept became commercially available with the Livescribe pen. The *ChronoViz* system [164] further refined this concept to support data collection and analysis in observational research. Handwritten field notes are automatically linked to audio and video recordings made during the observation and serve as bookmarks for subsequent data analysis.

The *PaperPDA* [35] is one of the early augmented paper notebook systems. It demonstrates how rich paper-based interactions can be realized even without digital pens, only by using a standard desktop scanner. PaperPDA combines a conventional paper notebook, calendar and organizer with electronic support. Using specific pre-printed paper forms, the user can for instance take notes and write e-mails (Fig. 2.12). This is particularly helpful in mobile settings. Back in the office, the

2.2 Pen-and-Paper Interfaces

Fig. 2.12 The PaperPDA. Center: A form for writing e-mails on paper. Right: Physical stickers for creating hyperlinks (photo courtesy of Scott Hudson)

user scans all new or updated forms using a desktop scanner. Specific visual marks on each form allow the system to identify the orientation and the type of the form. PaperPDA then detects marks that the user has made on interface elements (such as printed check boxes) and uses optical character recognition for recognizing text. The PaperPDA system then automatically performs the electronic operations requested by the user, e.g. sending an e-mail. Hyperlinks between two paper documents can be created with physical stickers (Fig. 2.12 right). The user attaches two corresponding stickers to the locations of both link anchors. On each of these stickers, an ID is printed that enables the system to detect corresponding stickers when the forms are scanned. This automatically creates a digital version of the hyperlink.

Collaborative Notetaking

While earlier systems mostly addressed the integration of digital media into the augmented notebook, recent research focuses on supporting collaborative uses of notebooks.

A good example of co-located collaborative use of paper notebooks is *AirTransNote* [97], a system for classrooms. Students take handwritten notes on a paper notebook. The notes are digitally captured and automatically transmitted to the PC of the instructor. This enables the instructor to provide feedback and to discuss notes of students by projecting them onto an electronic whiteboard. Moreover, the system uses handwriting recognition and clusters the recognized notes of all students. This allows the instructor to quickly get an overview of the students' notes.

Remote sharing of notebook contents via a network connection is supported by a number of systems. In the most simple case, the entire notebook is made available to other users without the option to select individual contents to share. This is the case with Memento [168] which allows for accessing shared notebooks in a standard Web browser. The *iDeas* notebook [69] (Fig. 2.11), a collaborative version of ButterflyNet, offers similar functionality. However, it also lets users select specific contents of their personal notebooks to be added to a shared group notebook. Maldonaldo et al. [89] have conducted one of the very few long-term studies of Pen-and-Paper Interfaces. Over a period of 6 months, more than 50 design students used the iDeas notebook for their studies. Main findings were that approximately two third of the participants adopted the novel technology and regularly used the system. While the participants disliked the large size of the pen and its short battery lifespan, they did not create less notes than students using a traditional notebook. Main advantages were seen in sharing notes with other team members and in automatically having a digital copy of the notebook. Prism [149] (introduced above) enables users to share individual pages of their paper notebooks with collaborators using Atom feeds. Sharing is controlled via a Web interface, but not directly from paper. Recently, *UbiSketch* [20] demonstrated how paper notebooks can be integrated with social media. Users can easily share their paper-based sketches on Facebook, Twitter and via e-mail, even on-the-go by connecting the digital pen to a smart phone. Results from a four-week user study indicate that shared sketches stimulate a higher degree of social interaction than shared photos. *PaperSketch* [166] uses a similar concept for sharing sketches with remote users in real-time via a Skype connection. Simultaneously users can communicate over voice and video channels.

EdFest [134, p. 153 sqq.] is a particularly rich augmented notebook that integrates notetaking with printed information and audio feedback. EdFest is a mobile interactive festival guide in a notebook format. The guide contains information about events of a festival, including their title, time, location and a short textual description. With an Anoto pen, the user can write handwritten reviews of events. These notes are transmitted to a central sever. Another user requesting information on that specific event can then access this review via text-to-speech output on a headset. This supports true mobile use, not only for data input, but also for output. EdFest provides additional functionality for a variety of festival-related tasks, such as rating events and getting the directions to an event location.

2.2.4 Augmented Printed Documents

Paper notebooks have the property to be initially empty. It is the user who, bit by bit, fills the notebook with handwritten contents. Yet, people do not only make handwritten notes on empty sheets of paper, but also write on paper documents that contain pre-printed contents. For instance, people use pen and paper to fill in questionnaires and forms, to make annotations on printed documents, to mark up contents or to create references. Such activities enable successful active reading, but also support effective presentation and discussion. The conceptual difference to augmented notebook systems is that handwritten contents must be matched with already existing printed contents. In this section we review a class of systems that allow users to create handwritten annotations on printed documents – annotations ranging from the highly structured answers written into a printed form to highly unstructured free-form comments.

Form Filling

Printed paper forms are still widely used today. They are inexpensive to produce and can be easily used at varying places. Paper forms are also the method of choice when computers are not appropriate for entering data, e.g. because they would create an interactional barrier, such as during interviews or medical consultations. However, most data collected in paper forms has eventually to be integrated into computer databases. Manually copying the data to a computer system or scanning paper forms requires additional effort and increases the time span until the data is electronically available.

This is where form-filling approaches have their benefits. They support structured data entry in mobile settings, retaining the advantages of paper forms. A paper form contains printed fields that act as placeholders where the user can fill in the requested data with a digital pen. This pen automatically captures all data and sends it to a computer system, e.g. via a mobile phone to the back-end system of a company, where the data is further processed. Figure 2.13 shows an example of a form-based interface.

The main commercial supplier of such solutions is Anoto, together with a number of partner companies. Form-based solutions are successfully used in a variety of settings.[20] For instance, field staff of several large companies use form-based solutions for filling-in order forms during their customer visits. Policemen use them

Fig. 2.13 Form-based interface (photo copyright Anoto)

---

[20] http://www.anoto.com/all-cases-4.aspx

for recording the details of traffic violations. Physicians and nurses take medical records in hospitals. The technology was even tested in the course of German elections [161], but eventually not used in subsequent elections due to security problems in the electronic voting system. More advanced solutions, that go beyond simply recording the data entered in a form, are currently investigated in research projects. For instance, one project examines how printed forms can support speech-language therapy [117].

Annotating Printed Documents

Capturing handwritten annotations on a printed document and synchronizing them with a digital version of the document is one of the most frequently addressed issues in Pen-and-Paper Computing. A very influential system is *Paper Augmented Digital Documents (PADD)* [32]. This system introduced the main principle of paper-based annotation: A digital document is printed on paper. This printout is used as a proxy to interact with the digital version of this document. Handwritten annotations made on the printout are automatically digitized and added to the digital document, which can be accessed on a computer. Moreover, the user can print updated versions of the document that include new annotations. This results in a digital-paper-digital annotation lifecycle. The PADD prototype uses an Anoto pen comes with a plug-in for Adobe Acrobat to annotate PDF documents. This concept was taken on by commercial solutions, including Anoto PenDocuments Pro[21] and Adapx Capturx Markup for PDF[22].

Proofrite [18] is a paper-augmented word processor that applies the PADD concept to the use case of proofreading. The user can annotate the printed version of a word processor document with proofreading marks that are synchronized with the digital version. In the digital view they reflow automatically when the surrounding text is modified. Neither PADD nor Proofrite do interpret the handwritings, but only capture and visualize them as an additional layer on top of the digital document. *PaperProof* [165] demonstrates how handwritten proofreading marks can be automatically interpreted to modify the underlying digital document. In contrast to the above systems, PaperProof creates a detailed semantic model of the printed document that captures what contents are printed at what locations on the paper document. This allows PaperProof to relate proofreading marks to document contents. For instance, if the user crosses out a word, PaperProof recognizes this gesture and identifies which word of the textual document is printed at the position of the cross-out mark. The word is then deleted in the word processor (see Fig. 2.14). Different gestures support inserting or moving text and making annotations. The prototype is integrated into OpenOffice.

The PADD principle was not only applied to word processing and standard office document formats. For instance, Pen-and-Paper Interfaces allow for annotations on

---

[21] http://www.anoto.com

[22] http://www.capturx.com

2.2 Pen-and-Paper Interfaces

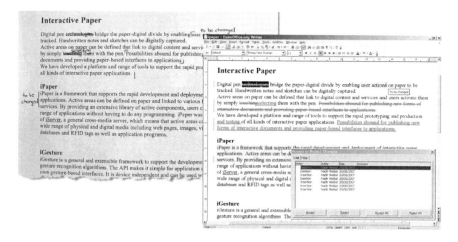

Fig. 2.14 PaperProof automatically interprets proofreading marks made on a printout (left) and applies them to the document in the word processor (right) (photo courtesy of Nadir Weibel)

large maps in geographic information systems [94]. Another example, *Musink*, supports annotations on musical partitions [154]. The system automatically analyzes them and integrates the printed partitions with OpenMusic, a computer-based music composition tool.

ModelCraft demonstrates that pen-and-paper annotations are not restricted to flat paper. The system supports handwritten annotations on physical 3D models that are made of paper [140]. Figure 2.15 shows a 3D model with some annotations. The annotations are automatically added to the underlying CAD model. However, in contrast to flat documents which can be easily re-printed, creating updated versions of physical models requires manual work. This limits the paper-digital annotation lifecycle.

One important benefit of digitized annotations is that they are not bound to the physical medium they have originally been made on. Hence, they can be more easily shared with other people, either in co-located settings or over distance:

Fig. 2.15 ModelCraft provides for annotating three-dimensional paper models (photo courtesy of Hyunyoung Song)

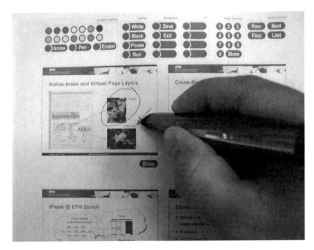

Fig. 2.16 PaperPoint integrates paper-based annotations into slide presentations (photo courtesy of Beat Signer)

PaperPoint [135] shows how a paper-based interface can support slide presentations. The presenter uses a printout of the presentation slides (Fig. 2.16). By making pen gestures on the printout, the presenter can control which slide is presented. In addition, handwritten annotations made on the printout are automatically integrated into the projected slide in real-time. The concept allows presenters to overcome the implicit linear character of slide presentations by developing a slide sequence on-the-fly that meets the demands of the audience. Anoto has commercialized this concept as a product called Anoto PenPresenter[23].

Similarly, *PaperCP* [74] supports in-classroom collaboration by slide annotations. Students can annotate printed handouts of lecture slides and electronically transmit their annotations to the instructor. The authors improve over PaperPoint by introducing a paper-based mechanism for defining which annotations are shared with the instructor and which annotations are kept private. This feature relies on two separate areas on each slide – one is used for making shared notes, the other for private notes.

CoScribe (cf. chapter 5 of this book and [144]) supports asynchronous sharing of annotations made on printed PDF documents and presentation slides. Different personal annotation styles are addressed by flexible print layouts, e.g. additional notetaking areas support extensive notes or many shared annotations. A paper-based sharing mechanism allows for sharing annotations with different groups of people. A novel visualization of shared handwritten annotations provides efficient access to annotations of multiple users in one single view.

---

[23] http://www.anoto.com/facilisis-a-consequat-quis-1.aspx

2.2 Pen-and-Paper Interfaces

Handwritten Categorization of Contents

Tagging is a more formal type of annotation. A tag is an annotation made on a document that describes its contents in a very condensed form, e.g. by a keyword. The goal of tagging is to classify documents and passages of documents in order to retrieve them more easily in the future. On the Web, tagging of resources is nowadays very common. Browser bookmarks allow users to make their personal bookmarks. Collaborative bookmarking tools such as del.icio.us[24], digg[25] or cite-u-like[26] enable sharing bookmarks with other users. Here we review concepts for tagging contents that are printed on paper. Most previous work relies on pen gestures.

PapierCraft [75] extends the PADD system for paper-based annotations, which was introduced above. It is a gesture-based command system for paper documents, relying on the Anoto technology and containing a rich set of pen gestures. These gestures enable not only tagging documents, but also creating and following hyperlinks on documents, copy/paste within paper documents, performing Google searches using keywords from the printed document and mailing portions of a document to other persons. To perform a command, the user must first select the portion of the document that the command applies to: Text snippets can be selected by underlining or encircling; several lines of text can be selected with a vertical line drawn besides the lines in the margin of the document; also arbitrary rectangular areas can be selected. Then the command is specified with a specific gesture.

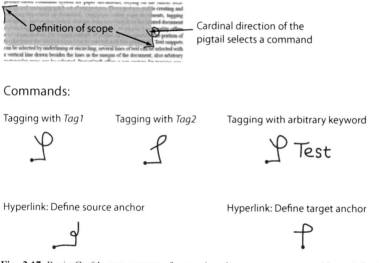

Fig. 2.17 PapierCraft's pen gestures for tagging document passages with predefined or freely-chosen keywords and for creating hyperlinks

---

[24] http://del.icio.us
[25] http://digg.com
[26] http://www.citeulike.org

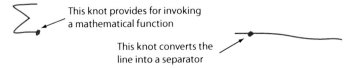

Fig. 2.18 Two examples of Knotty Gestures

As an example, Figure 2.17 depicts PapierCraft's gestures for tagging and linking contents. PapierCraft offers to select one of two predefined tagging categories. The cardinal direction of the ending of the pigtail gesture decides upon the category. This approach guarantees readability on paper and a reliable gesture recognition, but is restricted by the small number of categories that can be supported (up to the eight cardinal and secondary directions). Moreover, the abstract gestures have no natural connection to the category. As an alternative, the user can also create tags with an arbitrary handwritten keyword. To allow the system to reliably detect whether the user is writing or performing a gesture for invoking a command, the authors suggest using a secondary device, such as a push button.

Schumacher et al. [127] demonstrate how handwritten keywords can be further automatically processed. Similar to PapierCraft, their concept allows the user to select a passage of a document and to define a handwritten keyword. The system automatically recognizes the handwritten keyword and also extracts the selected text. This information is stored in the user's personal information base, an RDF-based ontology, and can henceforth be used for queries.

Knotty Gestures [155] aims at overcoming some of the limitations of traditional pen gestures. Handwritten gestures, such as those of PapierCraft, clutter up the content and take valuable space on paper if they are used heavily. Knotty gestures are less obtrusive. A knotty gesture is a small knot that resides on top of any other pen trace (see Fig. 2.18). One or several knots can be added at arbitrary points on an existing trace. The knot defines the role of the trace on which is resides. For instance, a knot can convert a line, which per se has no clearly defined semantic meaning, into a link anchor, a separator or an interactive slider. As another example, knots can be used for controlling recording and playback of audio recordings. Moreover, knots are not only useful during their creation; they also define a point of interaction that can be revisited at a later point in time. For instance, tapping with the pen on an already existing knot triggers associated commands. Circling around the center of a knot gives access to a list of options and selects one of them. The system is prototypically implemented on a Livescibe pen. The authors show that gestures can be reliably recognized directly on the pen, without the need for an additional computing device.

CoScribe (cf. Chapter 7 of this book and [144]) introduces a set of techniques for tagging contents on paper documents that goes beyond pen gestures. Digital Paper Bookmarks are physical stickers that can be attached to pages of a physical document for tagging them. The position of the bookmark and its handwritten label are automatically captured and made available in several visualizations. Bookmarks can be shared with other users over a network connection and can be compared in a

2.2 Pen-and-Paper Interfaces

collaborative view. A second tagging technique uses separate paper cards containing an inventory of tagging categories that can be modified and extended by end-users.

Handwritten Referencing

Handwritten annotations often contain references to different passages of the same document or to a second documents. Such user-defined references are not only beneficial for quick retrieval of information but also help in integrating and structuring information from many sources. We will now review pen-and-paper systems that provide for creating and following hyperlinks on paper.

PaperLink [5] was an early system within this class. It uses a specific pen onto which a camera is attached (Fig. 2.19 shows a commercialized version of Paper-Link). This camera serves for creating and detecting link hot-spots. The visual contents of the document serve as link anchors. If the pen is placed on a paper document, the camera captures an image of the document area around the pen tip. The image is processed on a computer using simple computer vision techniques. The pattern which appears at the center of the image (typically an individual word) is extracted. This pattern can be associated with a digital resource and serves from now on as a link hot-spot. If it is detected in the camera image, the target resource is opened on the computer.

The *Interactive Multimedia Textbook* [68] offers similar pen-based interactions for creating and following hyperlinks from printed documents to Web pages. It relies on an ultrasonic pen. Therefore link anchors do not have to be bound to specific

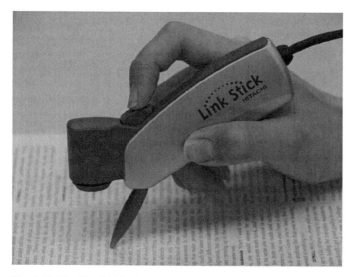

Fig. 2.19 The Hitachi LinkStick is a commercialized version of PaperLink. The pen has a built-in camera. From the camera image, a text pattern is extracted and acts as anchor for a hyperlink (photo courtesy of Toshifumi Arai)

visible marks on the document (e.g. an individual word). Instead hot-spot areas can be freely defined on the paper surface. Tapping with the pen on such an area displays the associated digital document. Both PaperLink and the Interactive Multimedia Textbook are limited to links from paper to digital media.

PapierCraft [75] (introduced above) allows the user to create hyperlinks between two paper documents. A hyperlink is created by drawing two specific pen gestures on both link anchors (see Fig. 2.17 on p. 53). To create links from or to digital documents, the same gestures can be used on a Tablet PC on which the digital document is displayed. In contrast to the other solutions discussed in this section, it is not possible to activate a hyperlink on paper, but only in the PapierCraft software viewer.

The hyperlinking functionality of *CoScribe* (cf. Chapter 6 of this book and [144]) focuses on tasks that require integrating and structuring information from a variety of documents, including printed documents and Web pages. Hyperlinks are created with pen gestures similar to PapierCraft's gestures. CoScribe introduces the following two novel aspects: On the one hand, it integrates the linking interactions; the same Anoto pen and the same gestures can be used both on paper and on displays. This avoids the need for switching between different devices and different interaction techniques, allowing seamless linking between both realms. A plug-in for Mozilla Firefox supports pen-based linking within Web pages. On the other hand, CoScribe automatically integrates linking activities of multiple users and multiple documents into one workgroup view. This view allows users to identify higher-level linking patterns that go beyond individual links.

In-place Visual Feedback

Almost all of the solutions discussed above strongly separate paper-based input from digital output. Digital versions of the documents are made available on a separate screen. We conclude this section by reviewing several publications that examine how digital information can be visualized directly *within* the paper document.

PenLight [138] introduced the concept of a digital pen that features a built-in projector. The pen projects a superimposed layer of digital information onto the paper document. This layer shows additional contents, such as annotations made by other users or additional images, but also provides visual guidance for invoking commands from pie menus. As current mobile projectors are still too large to be mounted to a pen, the authors simulat a projecting pen by tracking the position and orientation of the pen. Contents are then projected by a projector that is mounted at the ceiling above the table. *MouseLight* [139] is a follow-up work by the same authors. Here a real mobile projector is used, which however is detached from the pen and forms a separate device. This device, which resembles a mouse, can be placed onto paper documents and projects a superimposed layer of digital information. Figure 2.20 depicts MouseLight. The concept is similar to PenLight, but interaction is now bimanual. One hand is interacting with the pen while the other hand is manip-

2.2 Pen-and-Paper Interfaces

Fig. 2.20 MouseLight combines a digital pen with a separate mobile projector (photo courtesy of Hyunyoung Song)

ulating the projecting device. This leaves more interactional flexibility and ensures a stable projection even while the pen is used for writing.

Similar to MouseLight, *FACT* [78] uses a pen and a separate projection unit. In contrast to MouseLight, FACT leverages a camera and analyzes visual features to automatically identify paper documents and to track their position and orientation in 2D space. As a consequence, no Anoto pattern is required and documents can be placed and moved quite naturally on a table. The projector provides digital output on the paper documents in real time, enabling computer-like functionality on paper. For instance the user can select a word for performing a full-text search by tapping with the pen on a word that is printed on the document. All occurrences of this word are then highlighted by the projector directly within the printed document. Moreover, the system projects digital annotations onto the paper document, such as links to related Web pages. Further functionalities include Web search, copy/paste of paper contents and remote sharing. PACER [77], a predecessor system of FACT, enables similar interactions with paper and a mobile phone, however without in situ projection of digital contents.

2.2.5 Augmented Tables, Flipcharts and Whiteboards

Augmented tables, flipcharts and whiteboards combine paper-based media with interactive tabletop and wall displays. They provide a close integration of paper and digital media by combining them on one single surface. Paper documents can be used directly on top or in front of a large computer display. Alternatively, or in addition, digital information is displayed directly on paper.

Above we have already reviewed some concepts to display digital information in-situ, within a paper document, e.g. by leveraging mobile projectors or superimposed PDA displays. However, digital information was restricted to a small surface. In contrast, augmented tables, flipcharts and whiteboards realize very large digital display surfaces. These systems combine many of the augmented paper principles that we have introduced above.

Early works on augmented desks date back many decades. Already in 1945, Vannevar Bush envisioned *Memex* [12], an augmented desk that is able to display virtual documents on the table surface. Memex was thought of as an electro-mechanical machine that stores all books and other documents of a user on microfilm. Memex enables the user to read and annotate these documents as well as to create associations (we would call them hyperlinks today) between these documents. Memex never went beyond a theoretical state; however, it strongly influenced the development of hypertext systems and of augmented desks.

The foundations of digital augmented desks and tables, and of paper-based computing in general, were laid in a series of seminal works at EuroPARC in the early 1990s. A highly influential system is Wellner's *DigitalDesk* [167]. As shown in Fig. 2.21, the user can place printed documents on the table, very much like on a traditional desk. In addition, digital documents are projected onto the desk surface, and printed documents can be overlaid with projected information. This creates a very seamless integration of physical and digital contents.

The combined tracking and projection setup of the DigitalDesk inspired many follow-up works. It is depicted in Fig. 2.21 (left). A camera is mounted at a fix position above the desk. By analyzing the stream of images captured by this camera, the system identifies the position and the contents of printed documents. Moreover, the camera images are used for detecting pen and touch input on the desk – both on paper documents and on projected contents. Similar to the camera, the projector is mounted at a fix location above the desk.

Several example applications of the DigitalDesk introduced a set of novel interaction techniques. One interaction allows for physical-digital copy&paste of data. In a calculator application (Fig. 2.21 right), the user can copy numbers from printed documents by pointing on them. For recognizing numbers within the document snapshot captured by the camera, optical character recognition is used. Another interaction realizes physical-digital copy&paste of images. The user can draw a sketch on a sheet of paper and copy this sketch to a digital collage using a simple gesture. In the background, the system extracts an image of the sketch from the camera snapshot and projects this as a part of the collage. An extension of the initial system [124] in addition allows users to follow hyperlinks on printed versions of Web pages. By

2.2 Pen-and-Paper Interfaces

Fig. 2.21 The DigitalDesk. Left: Tracking-projection setup of the prototype. Right: copy&paste between a paper document and a calculator application (photos courtesy of Pierre Wellner)

tapping with the pen on a printed hyperlink, the target Web page is displayed in in a browser window that is projected onto the desk besides the paper document.

The DigitalDesk was the conceptual basis for a number of subsequent systems developed at EuroPARC. The *Digital Drawing Board* [86] is both a conventional drawing board and a top-projection display. It allows designers to make paper-based construction sketches on the board and to easily digitize them. The digitized images can then serve for a variety of purposes. For instance, they can be automatically rendered as textures onto objects to ease comparison of alternative designs.

Video Mosaic [85] addresses video editing. It combines paper video storyboards with the capabilities of video editing software. Obviously, planning and creating the temporal sequence of scenes is at the heart of video editing. The work examines how to best support the user in seeing in one glance such dynamic data. The chosen solution is to lay out time in physical space. Paper-based storyboard elements act as proxies for video clips. By physically arranging storyboard elements in a sequence on the desk, the user can change the sequence of clips to be played. Each storyboard element can be annotated with handwritten notes and sketches and can be associated with additional digital resources. The system allows the user to easily play back the current sequence, to digitize and reprint the storyboard elements, as well as to share them with co-workers.

EnhancedDesk [64] is based on a similar setup as the DigitalDesk, but uses two cameras for tracking. It further improved hand tracking and gesture recognition.

Fig. 2.22 The Shared Design Space integrates paper with an interactive tabletop and an interactive wall in a collaborative environment (photo courtesy of Michael Haller)

Digital documents are identified by fiducials. In an example application, a physical textbook is augmented by additional digital information that is projected onto the desk surface besides the textbook.

The *Augmented Surfaces* system [123] leverages both tables and walls as shared information surfaces. Using a mouse cursor, information can be seamlessly moved between these surfaces and ordinary computer displays. Moreover, the system leverages fiducials for identifying physical objects that are placed on the table. While the DigitalDesk provides for copy&paste from paper to digital, Augmented Surfaces support the reverse direction: digital contents can be virtually attached to a physical object. These digital contents are projected either directly onto the object or onto the table around the object, enabling compound collections of physical and digital objects. When the user moves or rotates the object, the projection is automatically updated to create the experience that the projected contents are physically attached to the object. A number of further seminal tabletop systems, such as *metaDESK* [156], were introduced in the late 1990s. These are not discussed here because they do not offer Pen-and-Paper Interaction.

The *Shared Design Space* [33] (Fig. 2.22) integrates paper with tabletop and wall displays. Similar to Augmented Surfaces, the user can move digital contents (e.g. videos and images) onto pages of paper notebooks. Moreover, the system automatically captures annotations and sketches that are made in the notebook with an Anoto pen. A clone functionality synchronizes notebook pages between multiple users in real time, blurring the boundary between the private notebook and the shared space. While the user who is making an annotation is leaving physical ink traces on his or her notebook, the collaborators simultaneously get a virtual copy projected on their own paper. Similar in approach, *Pictionaire* [34] is an augmented tabletop for collaborative design brainstorming. It coherently integrates the hybrid copy&paste

2.2 Pen-and-Paper Interfaces

operations introduced by the DigitalDesk and Augmented Surfaces. A drag-off gesture creates a digital copy of sketches that were made on physical paper. In the reverse direction, the user can snap digital images to paper surfaces.

Some time after interactive desk and table systems allowed for combining paper and displays in a horizontal configuration, interactive wall systems addressed the use of paper directly on vertical displays. The *Designers' Outpost* [63] focuses on creative planning tasks with post-it stickers (Fig. 2.23). It features a rear-projection whiteboard on which users can create hybrid physical-digital collages. These collages consist of physical post-it stickers that are attached onto the whiteboard and of virtual pen traces that are made on the whiteboard. A set of interactions allows for attaching a post-it at an arbitrary location on the whiteboard, moving or removing it, as well as for adding virtual pen traces. The locations and contents of the post-it stickers are automatically captured by a camera. This allows the user to save all contents of the board with one simple click. Saved contents can be accessed either on the whiteboard or via a computer interface. *DigiPost* [50] uses a similar principle on a horizontal tabletop display.

Similar to Designers' Outpost, *DocuDesk* [27] captures the physical arrangement of paper items. Its aim is to support users in creating multi-way links between paper documents and digital documents that are displayed on a horizontal tabletop. The system automatically captures contents of paper documents as well as hyperlinks. This allows the user to quickly re-establish the state of open windows and documents when resuming a task. To do so, it suffices to place one of the documents onto the screen. All other documents which are linked to that one are then displayed on the screen.

Placing paper documents onto a display potentially occludes digital contents. Only recently, research started addressing paper-based occlusion of screen contents. An empirical study [146] analyzed spatial patterns of how printed and digital doc-

Fig. 2.23 The Designers' Outpost (photo courtesy of Scott Klemmer)

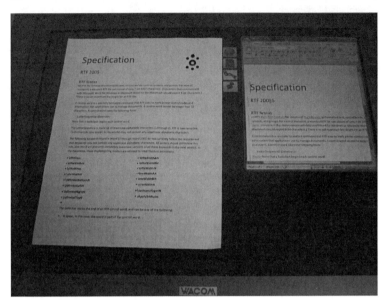

Fig. 2.24 DocuDesk (photo courtesy of Katherine Everitt)

uments are used simultaneously on the same surface and how users cope with occlusion. Another publication [57] introduced interaction techniques for managing hybrid paper-based and digital piles on tabletops that take into account paper-based occlusion.

Another strand of research examines how physical paper itself can become a large interactive surface, rather than combining (small) paper documents with (large) displays. The underlying premise of these works is that paper can be considered a display with a very slow update rate. A paper poster is therefore an inexpensive way of realizing a large, high-resolution display. *Gigapixel Prints* [177] presented a set of innovative applications of interactive paper posters. While input is realized with one or several Anoto pens, contents of the display can be updated either by reprinting the poster (slow, but high-resolution update) or by projecting additional contents onto the poster. The concept can play to its strengths when large data sets have to be visualized that are relatively stable over time. *PLink* [147] leverages large paper deskpads for integrating the physical desk and the computer desktop of office workers. The deskpad acts as a large linking area on which the user can easily and quickly create and access links to resources of the digital desktop. A four-week field study showed that PLink enables users to quickly access digital resources and supports flexible organization of digital information by laying out links in physical space.

The systems reviewed in the this section require that paper documents be kept flat on a two-dimensional surface. This restricts the natural interaction with paper, which to a large extent is manipulated above the table [151, 146]. Recent research opens up the 3D space above the table for paper-digital interaction:

2.2 Pen-and-Paper Interfaces

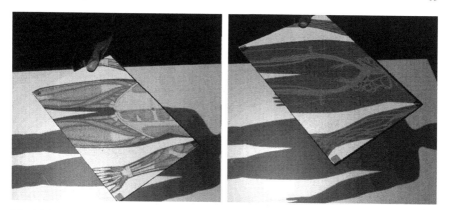

Fig. 2.25 PaperLens offers tangible views for information visualization above interactive tabletops (photo courtesy of Raimund Dachselt)

A very influential system is *PaperWindows* [40]. It introduces thin and lightweight paper displays as an interface for the windows of the operating system. Using a camera-projection unit, PaperWindows displays the contents of each GUI window onto one sheet of paper that the user can freely move on the desk. The authors present a set of interactions for copying windows from the computer to paper and back, for navigating through contents on the paper displays, for copy&paste of contents, and for annotating contents. These gestures leverage not only pen and touch input, but introduce paper manipulations, such as flipping or stacking pages, as a means for controlling the system.

PaperLens [142] (Fig. 2.25) is a tangible view for information visualization. A passive paper display can be used as an interactive lens for data that is displayed on an interactive tabletop. This is similar to the Interaction Lens of the A-Book (introduced above), but the PaperLens can be freely moved and rotated in three dimensions above the tabletop display. This allows for a set of novel interactions and visualizations that take into account not only the lens' position on the tabletop, but also its distance from the table surface and its 3D orientation.

The concepts discussed above use paper sheets of a fix size. One advantage of paper is that it can be modified in shape and size, for instance by folding. Lee et al. [70] examine different form factors of paper displays that allow for resizing the display. They present a large rectangular display that can be folded, a scroll that can be rolled in an out, as well as form factors for resizable displays which are inspired by fans and umbrellas. Inspired by this work, *Xpaaand* [56] addresses interaction techniques for such resizable displays. The authors present a hardware prototype of a passive rollable display and introduce a set of interaction techniques for manipulating digital information by resizing the display.

2.3 Directions of Future Research

This state of the art survey has shown that Pen-and-Paper Interfaces are technically mature and have found their way into commercially successful applications. A wide variety of technologies allow for capturing paper-based contents and for realizing digital output on paper. A large number of interface concepts cover three main functions of augmented paper with various paper media, ranging from tiny stickers over books and printed documents to large paper posters. The survey has also shown that a growing body of research aims at integrating paper more closely with real-time visual output, resulting in augmented digital pens and in paper-digital interactive surfaces.

To conclude this chapter, we briefly outline directions of future research on Pen-and-Paper Computing. A more comprehensive discussion can be found at the end of this book in Chapter 8.2. While most previous work has focused on data input on paper (output being provided on a separate computer screen), future work will aim at further *enhancing real-time feedback on paper*. Livescribe has shown the direction that future pens are likely to follow further: include more powerful processors to run complex applications directly on the pen and provide more real-time feedback on the pen. Future pens might feature larger displays, a built-in projector or even a built-in inkjet printer to leave permanent marks on paper. Moreover, mobile tracking-projection solutions are promising for transforming any paper document into an interactive surface. The increasing processing power of mobile phones and the advent of very small mobile projection units let us expect that in the near future, every smart phone comprises the components that are required for paper-digital interactive surfaces. A further, highly promising direction are flexible displays. These combine many affordances of paper with the powerful capabilities of displays. It is very likely that such displays will open up new and previously unforeseen ways of interacting with digital information.

On the level of applications, we see five major challenges. The first challenge is to fully leverage the *mobile character of paper*. Most prior applications either cannot be used at all in mobile settings or only parts of their functionality are available in a mobile setting. We expect more applications that couple paper and a digital pen with a mobile phone. This coupling results in a powerful device federation that requires only standard hardware which is already available today. Future work should examine how the user interface can be repartitioned between pen-and-paper and the mobile phone. A second challenge is related to how we manipulate paper in a Pen-and-Paper User Interface. Powerful new tracking technologies allow for capturing *manipulations that deform paper*, such as bending, folding and rolling. Future work should examine these interactions more deeply.

A third challenge consists of improving *large-scale collaboration*. Most current applications focus on a single user. In particular, it is still not fully understood how to process, integrate and visualize paper-based contents that are created by a very large community of users. This point is related to another challenge, the *interpretation* of contents. Almost all current systems interpret pen-and-paper interactions only to a very limited extent. The systems typically display only a facsimile of the handwrit-

2.3 Directions of Future Research

ten contents, sometimes performing handwriting recognition in the background to allow for full-text search. It will be interesting to see how contents that are created on pen-and-paper can be more directly integrated with tagging platforms, blogs and social networks, known as the Web 2.0.

This survey has shown that the hardware for realizing Pen-and-Paper Interfaces is readily available. However, the field lacks *standards and interoperability* of solutions. This concerns not only interfaces for abstracting from the pen hardware and an effective standard for digital ink data, but also standardizations in authoring and publishing.

Finally, we have seen that, even though a large number of Pen-and-Paper Interfaces was developed, there is only very little research examining *how people use these interfaces*. Most publications do either not report on user feedback at all or provide only limited insights. Only a small number of of studies examine how Pen-and-Paper Interfaces are used over a longer period of time and how they are integrated into existing information ecologies. Future work should definitely deepen our understanding and conduct more long-term studies of Pen-and-Paper Interfaces.

Chapter 3
Interaction Model of Pen-and-Paper User Interfaces

The advent of novel types of user interfaces generates challenges that relate both to the conceptual understanding and to the question of practical interface design. In this chapter, we aim at providing answers to the following questions:

- What is the essence of interaction with Pen-and-Paper User Interfaces?
- How to guide analysis?
- What aspects are to be considered when designing a Pen-and-Paper User Interface? What principles and guidelines do apply?

In essence these questions are answered for desktop interfaces (or GUIs), which have been the dominant interface paradigm for several decades. However, the large body of principles and guidelines established for desktop interfaces cannot be simply transferred to post-desktop user interfaces, of which Pen-and-Paper Interfaces are one form. While some first models of post-desktop interfaces are currently emerging, these do not take into account the specific characteristics of interacting with pen and paper. Hence they provide only limited support for designers of Pen-and-Paper User Interfaces.

This chapter provides the first theoretical model of interaction with Pen-and-Paper User Interfaces. An interaction model is a set of principles, rules and properties that guide the design of an interface [8]; examples of interaction models are the WIMP model and Direct Manipulation [133]. Our interaction model shows how to design paper-based interfaces that offer complex functionality, but that are nevertheless easy to learn and easy to use, easy to integrate in established contexts of use, and reliable despite the restrictions of pen and paper (e.g. limited feedback capabilities). The model takes on an integral viewpoint on the ensemble of collaborating users, of physical and digital artifacts, of work practices and of their interplay, which we consider an ecology.

We strongly believe that successful Pen-and-Paper User Interfaces are those that can be easily and seamlessly integrated into this ecology. This implies several requirements. Similar to established paper-based practices, user interfaces should heavily draw upon modularity and combination of tools. Taken on its own, each tool provides rather simple functionality which however can be used for variety of purposes (e.g. leaving ink traces on paper, leaving a bookmarking sticker, punching pages together). Depending on the user's needs and the current context, the user flexibly combines and/or repurposes these tools. Hence end-users becomes the designers of their interactions. This provides not only support for a variety of more complex tasks, but also allows end-users to integrate Pen-and-Paper User Interfaces with established tools at an equal footing, whether they be computerized or not. The model identifies a set of modular core interactions that are at the heart of pen-and-paper interfaces. Each of the core interactions is simple, easy to learn, easy to memorize and reliable. It is by the combination of several core interactions that more complex interfaces are designed.

We developed the model in an inductive-empirical process which generated theoretical results from an iterative user-centered design process (see Fig. 3.1). A theoretical process accompanied the user-centered design process. It provided theoretical input to the different phases of the design process. This input consisted of theoretical models of knowledge work and human-computer interaction, of empirical results from the literature and of design solutions from related work. In the reverse direction – from the design process towards theory – we subsequently abstracted the results of the design process in an inductive-empirical manner. This lead to the theoretical interaction model.

In this chapter, we start by defining Pen-and-Paper User Interfaces and briefly review related models. Then we present the theoretical perspective, which underlies the model, and discuss how to model interactions and information in Pen-and-Paper User Interfaces.

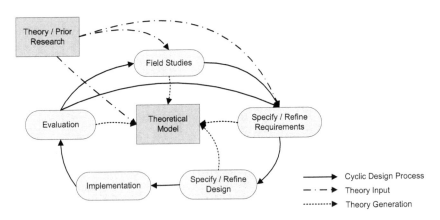

Fig. 3.1 Schematic overview of our research method

3.1 Pen-and-Paper User Interfaces (PPUIs)

One way to overcome the rupture between the paper world and the digital world are Pen-and-Paper User Interfaces (PPUI). PPUIs are part of the larger class of user interfaces which is called *post-desktop* or *post-WIMP* user interfaces. These interfaces go beyond the desktop metaphor and diverge from the "window, icon, menu, pointing device" (WIMP) paradigm of classical Graphical User Interfaces. Van Dam defines post-WIMP interfaces as "containing at least one interaction technique not dependent on classical 2D widgets such as menus and icons" [21]. These interfaces build "on users' pre-existing knowledge of the everyday, non-digital world to a much greater extent than before" [48].

PPUIs extend computing into the physical world by turning traditional paper into a digital interactive medium. Pen-and-Paper User Interfaces (PPUIs) consist of real paper and a pen whose movements are captured by the computer system. This enables for instance transferring handwritings and drawings to a computer and displaying a digital facsimile. Moreover, interactive elements of the user interface can be printed onto paper. For example, paper sheets can contain printed interface elements such as checkboxes, buttons, menus, fields for entering handwritten data or fields for issuing commands by drawing specific symbols. By interacting with a pen on these printed user interfaces, the user can control a digital system.

The basic setup of a PPUI is depicted in Figure 3.2. We distinguish two channels[1] that are indicated by the arrows.

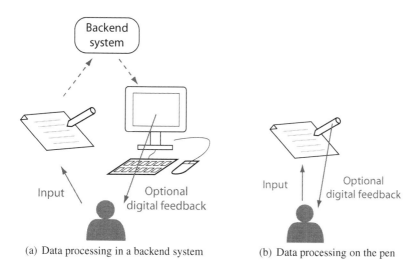

(a) Data processing in a backend system   (b) Data processing on the pen

Fig. 3.2 Basic setup of a Pen-and-Paper User Interface from a user's perspective

---

[1] A channel is realized by a physical device and transfers data from the user to the digital system (input channel) or vice versa (output channel).

In the setup depicted on the left hand side, digital pen and paper serve as a pure input channel to feed data into the digital system. Optionally, the system provides feedback via a separate channel. This is typically realized by a nearby display of a computer, a mobile phone, a PDA, or a projector, but the PPUI might also use auditory or haptic output channels. Hence, the PPUI can span several devices. The right hand side of the illustration depicts a different basic setup: Data is not processed on a backend system, but directly on the pen. Optionally, feedback is provided directly by the pen using a built-in display, audio feedback, a built-in projector or haptic output channels.

This basic setup considers only individual users. Yet, as we have discussed in Section 1.1, paper is a collaborative medium. Multiple physical objects in the physical space serve co-located interaction very well since they provide multiple points of interaction. Figure 3.3 depicts three settings that exemplify how several users can collaborate with one or multiple paper media. Paper media can be easily repositioned and combined, which provides for flexible collaborative settings. In a collaborative PPUI, several users can share the same paper interface or use different paper objects. Multiple pens can be used simultaneously, on a single or on multiple sheets of paper.

While paper is a very adequate medium for co-located collaboration, it is difficult to use paper for collaboration over distance. PPUIs can address this challenge and enable users to collaborate over distance not only by using *digital* but also by using *physical* documents. Two or more collaborators work at different locations and possibly at different points in time. Each user works on his or her local physical or digital representation of the document. Modifications that a person makes to a paper document are automatically distributed to his or her collaborators over a network connection and made digitally available (e.g. in a software viewer, by projecting changes onto the paper document or by reprinting the paper document).

Figure 3.4 depicts the collaborative setup of the Pen-and-Paper User Interface which we will present Chapters 4–7. Users can work individually, can share documents over a network connection and can cooperate in a co-located setup. The setup combines digital pens and paper with a PC, laptop or an interactive tabletop display.

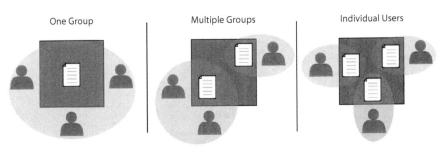

Fig. 3.3 Multiple points of interaction enable various co-located collaborative settings

3.2 Related Models

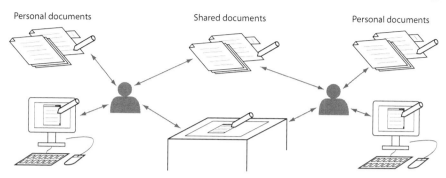

Fig. 3.4 The extended setup of Pen-and-Paper User Interfaces of our concept

3.2 Related Models

Almost all previous research on PPUIs focuses on interaction techniques and systems, but not on theory. In contrast, in the field of Tangible User Interfaces (TUIs) – of which PPUIs are a subclass – a growing number of publications address theoretical aspects. These concepts of TUIs can be used to theoretically describe partial aspects of PPUIs. We start by briefly reviewing research on TUIs and then discuss works on PPUIs that address theoretical aspects of interaction.

Theoretical frameworks of Tangible User Interfaces have made important contributions to the theoretical structuring of the domain by defining main components and conceptual terms as well as by categorizing systems. We can distinguish the following main dimensions which are addressed by these frameworks:

1. The coupling between physical and digital input and output
2. The conceptual and perceptual similarities between physical objects and their digital counterparts
3. The interaction with tangible objects in the physical space

Ullmer and Ishii [158] extend the well-known Model-View-Controller paradigm [67] to an interaction model for combined physical and digital interfaces. This describes elements for physical and digital input and output. It carries over the 'model' and 'control' elements while the 'view' element is divided into two subclasses. This accounts for the fact that most tangible interfaces represent the model's state by a combination of physical and digital information. The first subclass of the 'view' element are tangible objects which represent (parts of) the state of the digital model. These objects can also act as a physical control for the digital model. Second, a digital representation can provide further information on the system state, e.g. on dynamic information which might be hard to express by physical objects. For example, this digital representation can be provided on a nearby screen, via a speaker or by projecting a digital image onto the tangible objects.

Conceptual and perceptual similarities of physical and digital objects are addressed in Fishkin's taxonomy [29]. It includes two dimensions. The first dimension, which is called 'metaphor', focuses on the the perceptual similarity between physical objects and their digital counterparts. This similarity concerns both the physical shape/look/sound of the object and the actions that are performed with this object. The second axis ('embodiment') corresponds to the spatial distance between the tangible object, which is used for input by the user, and the device that provides system output. Fishkin distinguishes four levels of embodiment – ranging from feedback which is directly provided by the tangible to feedback which is provided on a distant device. Fishkin states that, in order to generate the impression of computation being embodied within the tangible objects, the tangible input device should also be the output device. However, in other situations, a spatially more distant system feedback might be more appropriate. Besides this spatial offset, Fishkin's taxonomy does not address temporal offsets. These occur in scenarios where physical interaction cannot be captured or processed in real-time by the computer system. An example is the situation in which a digital pen temporally buffers data when used in a mobile setting before being synchronized with a computer at a later point in time. The distance between input and output is also addressed by the framework of Koleva et al. [66], which denotes this as the 'degree of coherence'.

Other frameworks analyze the concrete interactions that the user performs with tangible interfaces. Interaction in most TUIs is centered on moving and arranging physical objects. For instance, the seminal URP system [159] enables urban planners to modify a digital model of urban buildings by rearranging physical models of these buildings. With other systems, such as the Marble Answering Machine [43] or MediaBlocks [157], the user accesses and modifies digital information by moving and arranging objects that act as physical handles for this information. Correspondingly, theoretical approaches to interaction within TUIs conceptualize interactions as changing the location or orientation of objects.

The framework of Ullmer and Ishii [158] classifies TUIs by the way in which they combine multiple tangible objects. The TAC paradigm [132] states that it is the physical constraints that define which interactions are possible (and not possible) with tangible objects in a TUI. Again, interactions are conceptualized as displacements and compositions of tangible objects. These concepts do not account for other types of interactions that alter the tangible objects themselves rather than displacing them. Ishii and Ullmer [43] transfer a set of GUI elements to TUIs (such as windows, icons and handles) suggesting generic physical instantiations of these elements. The focus is again on interaction as displacements, rotations and compositions of objects. Interactions with an individual object have meaning only with respect to other objects or to a reference frame. Finally, Wimmer [174] contributes a descriptive model of meaning that is expressed by grasping a tangible object.

All these theoretical frameworks do not account for the collaborative use of TUIs by multiple users. The framework of Hornecker and Buur [41] briefly discusses co-located use of TUIs by pointing out that TUIs offer multiple points of interaction, which provides for a spatially distributed control. Nevertheless, the collaborative dimension has not been extensively analyzed in TUI models.

3.2 Related Models

In contrast to general TUIs, research on Pen-and-Paper User Interfaces (PPUIs) almost exclusively focused on developing new interaction techniques and systems. A small number of publications took on a more theoretical point of view:

Guimbretière introduces a lifecycle model of transformations between paper and digital documents [32]. Although not presented as such, it is an important counterpart to Ullmer and Ishii's interaction model [158], which was mentioned above. Ullmer and Ishii model how *different* physical and digital representations are used simultaneously in a complementing manner (e.g. a digital projection overlaying physical building models). In contrast, Guimbretière models how the *same* document can be accessed in equivalent physical and digital representations. Depending on the situation, the user chooses the representations that best fit his or her needs. For example, in a mobile setting, the user might prefer reading and annotating a printout of the document, while he or she prefers working with a digital representation for sharing it with co-workers. Offering the user both a physical and a digital representation of the same information and letting the user choose between both of them is a dimension which, to our knowledge, is not considered in the research on general TUIs.

Yeh et al. [178] define a design space of paper interactions and present a toolkit for the rapid development of PPUIs. The toolkit offers generic elements for printed user interfaces, including input fields for handwritings and sketches, buttons and check boxes. However, the underlying (implicit) interaction model focuses on interactions with only single sheets of paper. It leaves aside the important dimension of physical arrangements of pages and of interactions that span multiple pages (such as Pick-and-Drop [122] and pen-based stitching gestures [37]).

Holman et al. [40] discuss how we might interact with documents on multiple digital paper displays, which are light, flat and malleable like paper. They introduce interaction primitives amongst others for activating and printing documents, for copy&paste and for scrolling within documents. These interactions rely on physical manipulations of paper displays, such as picking them up, collocating, flipping and stapling them, however not on pen-based interaction.

Finally, the iServer and iPaper framework [107] presents an extensive generic model for links between physical and digital documents. However, it does not cover specific interaction techniques.

This brief review showed that existing theory offers appropriate models for analyzing the interplay between physical and digital representations and for assessing the relations between physical objects and the digital information they represent. However, on the level of interactions, existing models do not sufficiently describe Pen-and-Paper User Interfaces. As discussed above, models of Tangible User Interfaces consider only the interactions of displacing, rotating and arranging physical objects while individual objects (e.g. shape, texture, content) are considered static. In contrast, interaction with pen and paper comprises writing with a pen. This implies that the tangible objects, i.e. the paper documents, themselves are altered.

An interaction model of Pen-and-Paper User Interfaces should identify generic interaction primitives that are performed with pen and paper. A first step towards these interaction primitives is the work by Yeh et al. [178]. However, this models only interactions with individual sheets of paper and not interactions that span multiple pages. As discussed in Section 1.1, main advantages of paper are precisely these multi-page interactions. In the following, we will present a generic interaction model that accounts for this dimension.

3.3 An Ecological Perspective of Document Work

Before modeling paper-based interactions and combined paper-and-digital information in the following sections, we first address what is the general setting we design for. Which general perspective should designers take when they design Pen-and-Paper User Interfaces?

Traditional cognitive approaches to human-computer interaction mainly focused on how an isolated user utilizes a computer system for performing an isolated task [125]. Transferred to our domain this could for instance mean that one designs a Pen-and-Paper User Interfaces that supports an individual user in one specific activity with one document, such as reading and commenting the document.

In this section we advocate a different perspective on document work. In our analyses of how university students use documents [145, 144, 143, p. 19 sqq.], it became obvious that working with documents was neither an individual activity nor restricted to using a single document. In contrast, students (and instructors) were in a permanent collaborative exchange, during and after courses. For instance students handed over notes taken during a lecture, jointly prepared for exams and discussed solutions of exercises. Very frequently, students concurrently used multiple documents in a tightly integrated way, for instance for combined reading and writing or for integrating knowledge from various document sources. This also involved simultaneous use of printed and digital documents. Similar findings have been reported in literature which analyzed other workplace settings (e.g. [131]). In summary, document work often takes place in a setting that encompasses multiple users and multiple documents. Individual activities occur in a complex network of users, documents, interleaving tasks and collaborative practices.

For this reason, a systemic viewpoint seems most appropriate for the design of PPUIs. We call this an *ecological perspective*. This perspective surpasses the view of individual users and individual tools. It is the integrated, systemic analysis of the elements of a particular knowledge work setting. This system consists of the *users*, of the *physical and digital artifacts and tools*, of the *practices* of using these artifacts and tools, of the *relations between users* as well of the *relations between artifacts and tools*. For instance, in our field of application, main elements of such an ecology are documents in physical or digital form, physical and digital tools that support working with these documents (e.g. pencils, rubbers, ring binders, digital pens, computers, mice, screens, printers and printed tools), users, practices as well as the relations between documents, between users, between documents and tools and between users and documents. These elements are depicted in Figure 3.5.

3.3 An Ecological Perspective of Document Work

Fig. 3.5 Main elements of an information ecology

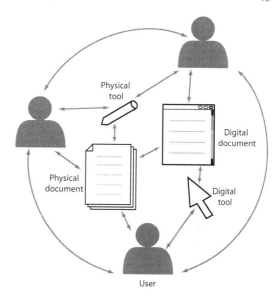

Now the question is: how do people perform cognitive activities in such an ecology? And how does the ecology deal with technological changes that are necessarily created by novel design solutions?

We base our discussion on two theories that provide answers to these questions: Distributed Cognition [42, 125, 39] and Information Ecologies [100]. Although both theories are quite different in their approach, they commonly advocate a systemic view focusing on the interrelations between actors, technology and given practices for understanding the use of technology in work settings. They argue that this perspective is the key for understanding and supporting knowledge work in a given context. Both theories provide different insights that are key to designing for document work.

Distributed Cognition is a theory about human cognitive processes. It provides a theoretical and methodological framework for analysis of collaborative work processes that include the use of technological artifacts and tools. The basic assumption is that cognition is not restricted to an individual nor restricted to the brain. Instead, cognition also occurs in the interactions between an individual and his or her environment (comprising other individuals and artifacts). Cognition is therefore embodied and situated within the work processes in which it occurs. Hence, the material world takes on a central rather than a peripheral role for cognition, as the work material becomes an element of the cognitive system itself. For instance, by arranging paper documents in a meaningful way on the desk, the user performs an embodied cognitive activity.

Cognitive activities take place in a 'functional system': "a collection of individuals and artifacts and their relations to each other in a particular work practice" [125]. Examples of functional systems include cockpits of airplanes, call centers and computer programmer teams. A cockpit, for instance, consists of pilots, of artifacts, such as instruments, displays and printed flight manuals, and of a set of practices, such as the precisely specified protocols for take-off or landing. In order to understand cognition in such a functional system, it is necessary to analyze the interaction of individuals, the ways they use artifacts and how all this is influenced by the environment. The focus is thereby on the question of how information is propagated across media, which means both internal media (the brain of an individual) and external media (e.g. a computer or a sheet of paper).

Distributed cognition emphasizes on the need of detailed analysis of such functional systems. As Rogers et al. [125] state, Distributed Cognition informs the design of computing systems for collaborative work by analyzing how novel systems might fit into current work practices and in which aspects they might be disruptive. Ethnographic studies of functional units are a central method for these analyses. For example, at first glance it might appear reasonable to share documents electronically within an organization. Yet, an ethnographic analysis of the functional system might for example show that e-mail is disruptive, as personally handing a physical copy over to a co-worker might fulfill a communicative purpose other than just passing the information contained within the document.

The theory of Information Ecologies adopts a similar viewpoint on knowledge work, even though not from a cognitive but from an anthropological perspective. Inspired from biological ecosystems, the authors introduce the idea that technology can be metaphorically seen as an ecology. They define an information ecology to be a "system of people, practices, values and technologies in a given local environment". Their attention is on relationships involving tools, people and their practices. Note that this is very similar to the functional system of Distributed Cognition. People correspond to individuals. Practices and values of a local environment correspond to a particular work practice. Technology corresponds to artifacts. Examples of information ecologies include libraries, self-service copy shops or intensive care units of hospitals. Like a biological ecosystem, an information ecology is a complex system of elements (comprising people and tools) which have strong interrelations and dependencies. It contains a diversity of roles for the people and functions for tools. In a self-service copy shop, for example, there are such various tools as copy machines, computers, scanners, paper stock and scissors. If customers need help on how to use a machine, they can ask one another or get helped by the staff.

The theory of information ecologies introduced two aspects which are of particular relevance for our domain: The first aspect is related to how an ecology deals with technological changes. The theory assumes that information ecologies are characterized by a continuous evolution. As novel technologies are integrated into current work practice, technologies and practices are adapted and mutually assimilated to fit to each other. Hence, when designing for an ecology, it is crucial to consider existing practices and to be aware that the novel system will be adapted – consciously or inconsciously – by the end-users to integrate into the (evolving) ecology. This directly relates to the second aspect, which the theory calls "locality". The same technology (e.g. the same type of computer with the same hardware and software configuration) can be used very differently in different environments. Hence, the local participants define the identity and place of the technology. As a consequence, it is the task of the designers to provide useful functionality, but the local participants complete the job by integrating them into their practices in a way that makes sense for them.

In summary, we derive the following implications for the design of Pen-and-Paper User Interfaces. First, the design should consider collaboration as a key aspect in document work. Second it should take the interdependencies between multiple (physical and digital) documents and (physical and digital) tools seriously into account. Third, detailed analysis of the ecology which to design for is crucial. Fourth, end-users adapt novel technology to fit into existing ecologies. Technological tools should be generic enough such that end-users can easily localize them. This general perspective allows us to address in the following more closely how interactions with Pen-and-Paper User Interfaces can be modeled.

3.4 Model of Interactions

Pen-and-Paper User Interfaces have other characteristics than Graphical User Interfaces (GUI), because they comprise physical objects as interface elements and use a digital pen as the main interaction device. In contrast, GUIs typically rely on keyboards and mice. Moreover, paper is a very restricted output channel, which makes it challenging to design a user interface that supports complex activities and still remains easy to use. This implies that it is not sufficient to transfer interface elements from the GUI to paper, such as transferring text input fields or buttons. We argue that instead, interactions should build upon specific paper affordances, such as simultaneously using multiple pages.

In this section, we present a model of pen-and-paper-based interaction which provides guidance for analysis and design of interfaces. The underlying principle of the model is an analytic separation of interaction into a semantic and a syntactic level (see Fig. 3.6):

1. The semantic level models *what* the user wants to do and comprises conceptual activities, i.e. the functionality offered by the user interface (for instance the activities of annotating, linking and tagging).
2. The syntactic level models *how* the user actually performs these activities. It comprises core interactions, i.e. primitive manipulations that are made with the PPUI in order to actually perform these conceptual activities (e.g. writing with the pen or attaching a paper sticker).[2]

The challenge when designing a PPUI is first to identify simple and reliable core interactions which leverage the affordances of pen and paper. Second, the designer must decide which core interactions to use and how the user combines them to

---
[2] This separation of interaction into two levels is conceptually similar to the four-level model by Foley et al. [30]. In addition to the semantic and syntactic levels, this covers a conceptual level, which is the user's mental model of the interactive system, and a lexical level, which encompasses the precise mechanisms by which the user specifies the syntax. We judge two levels to be sufficient for our aim of identifying core interactions and of modeling how these can be composed.

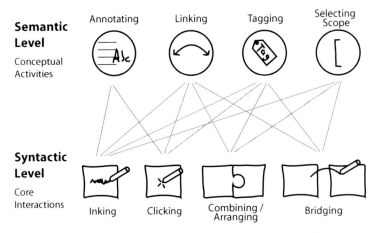

Fig. 3.6 The analytical framework (exemplarily applied to the activities supported by the CoScribe system, which is introduced in Chs. 4–7)

perform a conceptual activity. Ethnographic observations of users' current practices are an important method for informing these design decisions.

As we will show below in this section, this principle enables PPUIs that are both simple and complex: offering complex functionality while being easy to use. This is mainly due to the fact that the generic core interactions are quite intuitive, as they are inspired by the traditional cultural techniques of interacting with pen and paper.

3.4.1 Semantic Level of Interaction: Conceptual Activities

The semantic level of interaction models conceptual activities. This covers the functionality that is offered by the interface to support the user in his or her tasks. Hence, this level is not generic but depends on the purpose the PPUI is designed for.

As the goal of the model is to be generic, we do not go into depth of conceptual activities here, but illustrate the concept with one example, the CoScribe system which is introduced in Chapters 4–7. For our application scenario, we identified three main conceptual activities that users perform when working with documents:

1. Extending documents by annotations
2. Creating references and hyperlinks to express relations between documents and passages of documents and
3. Creating tags in order to structure the problem domain by translating given contents into higher-order concepts.

These conceptual activities are complemented by a fourth activity, which is an essential part of the other activities:

3.4 Model of Interactions

4. Defining the scope of an annotation, link or tag. This is the portion of the document (or of several documents) it applies to. In the literature, the scope of an annotation is often referred to as its context. The scope of a link is called its anchor.

Conceptual activities can be hierarchically organized. For instance, scope selection is a sub-activity of the three other activities. Creating a hyperlink includes the sub-activity of selecting the passage of a document the hyperlink applies to. We might also define higher-level activities. For example, the activity of "excerpting on a separate sheet of paper" relies both on annotation and on linking.

3.4.2 Syntactic Level of Interaction: Core Interactions

The syntactic level encompasses interaction primitives of PPUIs. These interaction primitives are independent of the functionality of the PPUI. Instead, they model in a generic way how users interact with pen and paper. Our interaction primitives are based on observations of pen-and-paper practice we made in several field studies [145, 144, 143, p. 19 sqq.]. We identified the following three main categories of interaction:

The first category consists of writing and drawing with a pen on paper (e.g. annotations, handwritten references, keywords or symbols for tags). People often partition the available space into separate zones for different functionality, e.g. reserving the left or right margins of the document for keywords in order to provide for a quick overview on all keywords on a page. A second category involves leveraging the material aspect of paper sheets, which can be flexibly moved and arranged in physical space. Specific spatial arrangements of two or more paper sheets convey semantics (e.g. relating documents by putting them into a folder or placing them on a stack, or marking important pages with bookmark stickers). Moreover, the shape of physical paper sheets can be modified, e.g. by bending or folding a sheet of paper, or by tearing it into several pieces. Finally, we frequently observed that people pointed to documents. This occurs most often in collaborative settings, but also when a single person reads on his or her own, for instance during intense reading or for temporarily marking a passage. Often people do not only point to one single document but consecutively point to several documents or several pages. This allows them to express relations between the contents they point to.

Based on these findings, we identified a set of core interactions, which are depicted in Fig. 3.7. A *core interaction* is defined as an operation that a user performs by manipulating one or more page areas using a digital pen or his or her hands. Examples of page areas comprise a printed document page, a printed button element or an adhesive paper sticker. First and foremost, page areas are contained on printed sheets of paper. However, by analogy, areas which are displayed on a screen can equally act as page areas. Note that the same core interaction can have a different meaning if it is performed on a different type of page area or using a different tool, e.g. a digital eraser. We distinguish the following core interactions:

80   3 Interaction Model of Pen-and-Paper User Interfaces

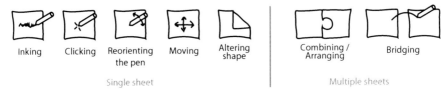

Fig. 3.7 Core Interactions of Pen-and-Paper User Interfaces

- *Inking:* Writing with the digital pen on a page area. This includes free-form handwritings and drawings that are digitally captured. Moreover, specific symbols and pen gestures can be performed to issue a command.
- *Clicking:* Performing one or more pen taps on a paper area to issue a command (e.g. on a printed "button" area). This is inspired by traditional pointing gestures. We distinguish clicking from inking for the following reason: While inking leaves visible pen traces and permanently alters the document, clicking is volatile, leaving it conceptually unchanged.
- *Reorienting the pen:* Modifying the spatial orientation of the pen, on or above paper. This comprises tilting and rotating the pen on the surface as well as pointing with the pen by hovering above the paper surface.
- *Moving:* Changing the physical location of the page area. This also includes picking it up and putting it down as well as flipping pages.
- *Altering shape:* Altering the physical shape of a page area, for example by bending, folding or tearing it.

An important characteristic of paper is that it affords using several sheets of paper at a time. The following core interactions comprise the combined use of two or more sheets:

- *Combining:* Creating or modifying arrangements of page areas. An arrangement can be rather volatile (e.g. paper sheets laid out on a desk) or rather permanent (e.g. attached paper stickers, documents filed in a folder, sheets stapled together).
- *Bridging:* In contrast to physical combinations, bridging is a logical combination of several areas. This complements physical combinations or substitutes them when these are impractical or impossible. Inspired by consecutive pointing on several items, we model bridging as a connecting pen gesture on two different areas.

PPUIs should account for the rich interactions that are possible with paper and use a broad spectrum of core interactions. In Section 1.1, we have seen that paper affords the flexible spatial organization and the concurrent use of multiple sheets. PPUIs should leverage this affordance and support the core interactions of combining and bridging. This stands in contrast to a design which is inspired by the GUI paradigm, in which interaction is restricted to single points of focus (due to the single mouse pointer).

Table 3.1 depicts how our core interactions relate to interactions with traditional paper and how they correspond to interactions within Graphical User Interfaces. As

3.4 Model of Interactions

Table 3.1 Comparison of Core Interactions

	Traditional Paper	PPUI	GUI (following [8])
Single sheet	Writing	**Inking**	Text entry
	Pointing	**Clicking**	Pointing/Clicking
	Pointing	**Reorienting the pen**	–
	Moving paper	**Moving**	Dragging
	Altering shape	**Altering shape**	– [a]
Multiple sheets	Arranging/Combining	**Combining**	– [a]
	Subsequent pointing	**Bridging**	– [a]

[a] No core interaction (performed by combining several core interactions)

GUIs incorporate metaphors of traditional desks, they offer somewhat equivalent interactions for inking, clicking and moving. We deliberately model combining and bridging as separate core interactions instead of combinations of inking, clicking and moving. The reason is that one important affordance of paper is to support two-handed interaction. Both hands can be used to simultaneously interact with two or more sheets of paper. This stands in contrast to traditional GUIs, which do not support simultaneous interaction at two or more focus points. Consequentially, combining and bridging in GUIs should be rather considered as sequential manipulations of individual areas, hence as a sequence of inking, moving and clicking. Finally, the interactions of reorienting the pen and altering the shape of paper leverage the specifics of a pen-and-paper environments and go beyond what is possible in classical GUIs (pointing, clicking and dragging with the mouse as well as resizing of windows).

To conclude the discussion of core interactions, we apply them to a representative set of PPUI systems that support users in interacting with documents. We demonstrate that these can be classified in terms of the generic core interactions identified above. Table 3.2 provides an overview of these systems.

A first class of systems (e.g. [32, 165, 135, 78]) augments paper documents by electronically capturing handwritten annotations (*inking*). ButterflyNet [176] additionally supports creating associations (*bridging*) between an area of a paper notebook and a digital photo with a pen gesture. Users can access the digital resource by tapping with the pen on this paper area. PapierCraft [75] supports tagging paper documents with pen gestures (*inking*). In a calculator application, Wellner's DigitalDesk [167] supports entering numbers by pointing (*clicking*) on a number in an arbitrary document on the desk, regardless if the document is printed or projected. Moreover, the position of digital and physical documents on the desk can be mirrored to collaborators over distance in real-time. Each time a physical or digital document is *moved*, the position is updated at the remote site. FACT [78] and PenLight [138] allow users to select printed content by *inking* gestures. In addition, PenLight leverages *reorienting the pen* in 3d space above paper. By hovering with the pen and moving it in 3d space, the user can select options from pie menus and moreover define 3d positions from which a perspective rendering of paper contents is calculated and projected.

Table 3.2 Core interactions of a representative set of related systems and of the CoScribe interaction strategies presented in this book

System	Inking	Clicking	Reorienting	Moving	Altering shape	Combining	Bridging	Purpose
PADD [32]	•							Capturing handwriting
PaperProof [165]	•							Capturing handwriting
PaperPoint [135]	•							Capturing handwriting
		•						Controlling slide presentations
ButterflyNet [176]	•							Capturing handwriting
						•		Linking digital data
DigitalDesk [167]		•						Drawing application
				•				Selecting numbers
				•				Moving shared documents
		•				•		Crating physical/ digital collages
PapierCraft [75]		•						Creating tags
		•					•	Creating hyperlinks
		•				•	•	Creating collages
FACT [78]		•	•					Selecting printed contents
		•						Selecting menu options
		•						Issue gesture-based commands
PenLight [138]		•						Capturing handwriting
		•						Selecting printed contents
			•					Selecting menu options
			•					Defining 3d perspective
CoScribe	•							Capturing handwriting
		•					•	Classifying annotations
		•						Following hyperlinks
		•						Controlling digital viewer
				•		•	•	Creating hyperlinks
				•		•	•	Creating tags

All these conceptual activities are performed with one single core interaction. However, there are also examples, where conceptual activities are supported by a combination of several core interactions. In PapierCraft, the user creates a hyperlink between two paper pages by first highlighting (*inking*) the passages that shall be linked and then *bridging* them with two consecutive markings on both pages. Moreover, PapierCraft supports creating physical collages: A user first physically *combines* two paper sheets by positioning one besides the other in a way that the margins slightly overlap. Drawing an associating line then digitally *bridges* both sheets. The Digital Desk allows users to select document snippets with a pen gesture (*inking*) and then *combine* these snippets in a physical and digital collage.

The novel interaction techniques presented in this book draw upon single core interactions as well as upon combinations of them. Table 3.2 includes an overview of these strategies, which will be discussed in the following chapters.

3.4.3 Mapping Between Syntax and Semantics

Having defined the semantic and the syntactic level of interaction, we now focus on the intersection between both levels. We analyze how syntactic core interactions can be mapped to semantic activities. The number of activities is larger than the number of core interactions. For this reason, the model must rely on some kind of multiplexing [8] between a smaller number of core interactions and a larger number of semantic activities.

We distinguish the following four types of multiplexing, which are used to map core interactions to activities. A syntactic interaction that performs a specific activity might use one or more of these types of multiplexing between core interactions and semantic activities. Figure 3.8 illustrates the four types of multiplexing.

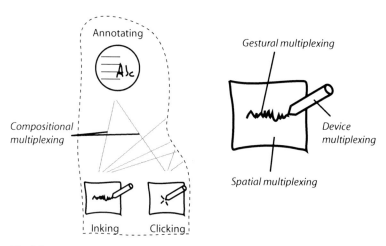

Fig. 3.8 Types of multiplexing between core interactions and semantic activities

Compositional Multiplexing A key finding of our field work was that in traditional paper practice, users often combine several core interactions to perform a single conceptual activity. The user for instance attaches an adhesive indexing sticker onto a document page and then writes a label on this sticker. We call this compositional multiplexing and judge it to be – together with spatial multiplexing – most important for interaction in PPUIs. This principle enables users to draw upon a small repository of simple core interactions. These act as flexible building blocks which are used and reused for multiple conceptual activities.

Spatial Multiplexing In traditional practice of working with paper documents, different sheets of paper or different areas on one sheet have different functions. For example, writing a label on an index sticker possibly has another meaning than writing the same label on a document page. Applied to our model of PPUIs this implies that PPUIs should comprise different types of page areas that serve different purposes (e.g. document pages, folders and index stickers). The same core interaction performed on a different type of page area leads to another result. For example, if the core interaction of inking is performed on an adhesive bookmarking sticker, it might define the label of an index, whereas inking on a document page might create a free-form annotation.

Spatial multiplexing results in a set of specialized tools made of paper. These tools have the characteristic that they are not only instruments but they can also become objects of interest. Tools that have a purely instrumental function are only used to manipulate objects of interest (e.g. documents). Examples of traditional instrumental tools are hammers, scissors and pens. Most tools in Graphical User Interfaces belong to this category. In contrast, we aim at designing tools that in addition to their instrumental function are first-class objects. For example, an adhesive sticker that can be attached to a document page in order to bookmark this page is initially an instrument for creating an index on a document page. But once the sticker is attached to a document page, it becomes also an object of interest, as it represents the actual index. Similarly, physical folders are used as an instrument to define collections of documents. In addition it is itself an object of interest. Its physical state indicates for example if only a few or many documents are included in this collection.

Gestural Multiplexing Different words or gestures written with the same pen on the same type of page area serve different purposes. For example the keyword "important" might serve the purpose of a tag while the term "cf. page 3" might serve the purpose of a link. Gestural multiplexing is very powerful, as a large number of gestures can be defined. Moreover, it does not require other tools than one pen and unspecific sheets of paper. However, a heavy use of gestural multiplexing runs the risk to create a command-based interface which requires that the user memorize a large number of commands and results in poor usability (see Norman & Nielsen's recent criticism of gestural interfaces [106]). A heavy use of gestural multiplexing does not account for the rich practice of traditional interaction with paper documents, which does comprise tools other than a pen, such as index stickers, page markers and folders. For this reason, we argue that Pen-and-Paper User Interfaces should use only a small set of simple gestures, such as points and lines, and put the emphasis rather on a rich variety of paper tools (and hence on spatial multiplexing).

Device Multiplexing If performed with another device, the same actions can have a different signification. For example, inking with a digital pen which has a ballpoint tip might be used for making textual annotations, whereas inking with another type of digital pen which features a highlighter tip might be used for marking up passages. Device multiplexing is powerful if the number of devices keeps manageable and if the repartition of functions across devices is clear. However, there is empirical evidence that people tend to use one single pen rather than switching between many tools [91]. As an alternative to switching between devices, the pen device can feature physical input elements that allow for switching between modes on one single pen.

3.5 Model of Information

A model of Pen-and-Paper User Interfaces should not only cover paper-based interactions, but also address how information is distributed between physical and digital representations. We distinguish two orthogonal principles that model the relations between printed and digital pieces of information on a semantic level: equivalence and complementarity. Our model draws inspiration from research on multimodality that formalizes the relationships between different modalities [19, 160].

Equivalence of Information We refer to the first principle as equivalence. It is illustrated in Figure 3.9 (left). This principle captures that the same piece of information can be transformed between equivalent representations on different media. The user can choose the representation which best fits her needs. As a matter of course, this is only possible if the piece of information disposes of an equivalent representation. For instance, a video document does not have an equivalent physical representation.[3] In contrast, a book, a printed article, or a PDF document have equivalent representations. Their contents can be transformed to a spatially fix, temporally static and two-dimensional layout that can be either displayed on a screen or printed on paper. No essential contents are lost and an unambiguous mapping from the printed to the corresponding digital contents is possible.

On the level of information, both representations are equivalent, as they contain the same contents. However, different representations can offer different affordances. For example, the user could utilize a printer to transform a digital representation to a printed one because it is more convenient to read and annotate information on paper. In the reverse direction, the user could transform a printed piece of information to an equivalent digital representation that better affords searching for specific terms.

Two or more equivalent representations of the same piece of information can be used one after another, or they can be used in combination at the same point of time. The latter combines the affordances of several representational media. For example the printed representation of a document can be used to quickly navigate between different pages of this document, while the digital representation better affords editing, moving or deleting existing annotations, as it is updated in real-time.

Complementarity of Information The second principle, called complementarity, models the orthogonal dimension. It is illustrated in Figure 3.9 (right). The information space consists of several, different pieces of information that are distributed across different representations and complement each other. Each piece of information is available in one fix representation. The distinctive features of both principles are shown in a matrix in Figure 3.10. In practice, both dimensions appear together. For example, one of the complementing pieces of information can in turn be transformed to an equivalent representation.

---

[3] Also a video document can actually be represented on paper (e.g. as a token containing only the tile or as a collection of printed key frames). However, this representation is significantly different from how a video is represented in a digital representation.

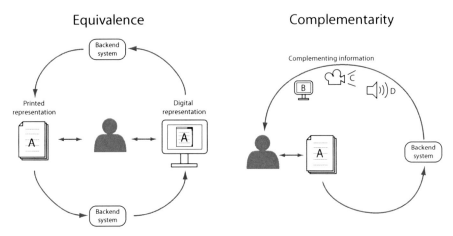

Fig. 3.9 The two principles of how information is related between different representations

	EQUIVALENCE	COMPLEMENTARITY
Piece of Information	**single** one single piece of information	**multiple** several complementing pieces of information
Representation	**multiple** multiple representations	**single** one single representation for each piece of information

Fig. 3.10 Distinctive features of the two principles

Complementing information serves two main purposes. First, complementing pieces of information can provide the user with updates of a piece of information. While it is usually preferable to directly update the representation of the piece of information the user is working with, this is not always possible. For example, a digital system cannot update information on paper in real-time. However, real-time updates are possible with a complementing digital representation, such as a projection that overlays the printed document.

Second, complementing information can provide different pieces of information on a second channel. For example, if the user works with a printed representation of a Web page which contains background music, a complementing digital channel

3.5 Model of Information

could play this music. A second example is system feedback, which can be given on another channel, i.e. in another representation.

Appropriate Complementing Representations In case of complementary information, which complementary modalities should be chosen? Table 3.3 provides an overview of combinations of paper-based and digital representations and of their properties. These can be classified by four dimensions:

1. The time needed for updating information in the complementing representation.
2. The spatial distance between paper-based and complementing representations. Ideally, contents of the complementing representation should be made available in-situ, directly at the location of the paper document.
3. Whether the same or a different modality is used (information that is printed on paper has a visual representation).
4. The amount of data that can be transferred using the complementing representation. The data rate should be high to provide comprehensive information (e.g. the contents of a document page), while representations with lower data rates can be used for purposes such as indicating success or failure of an operation or the current mode of the system.

Complementing information that is in situ, immediate, of the same modality and provides high data rates can be realized by a projection that overlays information that is printed on paper. Alternatively, a small and lightweight display can be placed onto paper in order to display additional contents. Both approaches require a rather complex technical setup that is not yet available as an out-of-the-box solution. A technically less complex solution for realizing in situ and immediate information with a high data rate consists of using an auditory channel. However, this channel has a different modality (which is not necessarily a drawback). While it offers a high data rate for conveying textual information, it cannot be used for transferring visual contents. If this restriction is not acceptable, extensive visual information can only be conveyed by accepting trade-offs with respect to distance or immediacy.

Table 3.3 Appropriate complementing representations

Representation	Complementing representation		immediate	in situ	same modality	high data rate
Paper	Visual	Overlaid projection	●	●	●	●
		Overlaid display	●	●	●	●
		Nearby display	●	–	●	●
		Re-print	–	●	●	●
		Pen display / LED	●	●	●	–
	Auditory	Speaker	●	●	–	●/–
	Haptic	Pen	●	●	–	–

On the one hand, information can be immediately provided on a nearby display, which however is not co-located with the paper medium. On the other hand, an updated printout of the paper representation can contain comprehensive updates and feedback, which is in situ, but delayed.

3.6 Conclusions and Design Guidelines

In this chapter, we presented an interaction model for Pen-and-Paper User Interfaces (PPUIs). The model provides a conceptual basis for the design of paper-based interaction techniques and systems. It was developed in an inductive empirical process and is grounded on findings from the literature, on our own field studies and on an analysis of existing user interfaces from related work. The model defines Pen-and-Paper User Interfaces, models interactions with PPUIs and models how information is distributed between paper and digital representations. The model implies the following design guidelines for Pen-and-Paper User Interfaces:

1. **Draw upon the affordances of paper and upon traditional practices of working with paper**
   Traditional practices of working with paper have evolved over a long period of time and have proven to be highly effective in many respects. As they leverage the affordances of paper, they have advantages that GUI-inspired interfaces do not necessarily provide (e.g. traditional interaction is reliable even though there is no real-time feedback provided by a computer system). In order to inspire the design of a paper-based application, the designer should use ethnographic methods to analyze what activities within the ecology the system should support (semantic level of the interaction model) and by which means users actually perform these activities (syntactic level).
2. **Design a modular interface using simple and flexible building blocks**
   The interaction model proposes a small inventory of generic core interactions. Each core interaction is inspired from traditional practices of interacting with paper and designed to be intuitive, simple to use and reliable. Using compositional multiplexing, core interactions can act as building blocks and can be flexibly combined. This provides for offering complex functionality in a PPUI which nevertheless remains easy to use, as only a very restricted number of core interactions is used on the syntactic level.
3. **Provide for rich interactions**
   The design should incorporate the richness of interacting with paper. First, this includes using a wide variety of core interactions, in particular those that combine multiple sheets (e.g. combining/arranging and bridging). This stands in contrast to a design that is inspired by the single point of focus of Graphical User Interfaces and that uses only the core interactions of inking and clicking. Second, PPUIs should include tangible tools that are made out of paper (e.g. leaf binders, index stickers). This leverages the power of spatial multiplexing, binding different functionality to different sheets of paper and thereby to different positions

in physical space. This results in more intuitive interfaces that moreover can be flexibly adapted to the current work situation by arranging tangible tools accordingly.

4. **Cope with the restricted feedback capabilities of paper**
A major challenge for the design of PPUIs is the very restricted feedback capabilities that we are facing when using paper as an interactive medium. First and foremost, the designer should identify interactions that leverage feedback which is provided by the physical properties of paper instead of digital real-time feedback which must be provided by the system. For instance, writing on a sheet of paper generates "real-time feedback" by leaving visible ink traces; attaching an index sticker generates feedback in form of a sticker which is visible on paper. Second, the design should avoid temporal modes if the current mode cannot be clearly and continuously communicated to the user. Instead, the design should leverage spatial multiplexing or device multiplexing – this provides for "modes" which the user can clearly recognize as such without digital feedback. Third, the design should make careful use of handwriting recognition and gesture recognition. These should be used as a central part of the interaction design only if the recognition does not imply uncertainty or if real-time feedback on the recognition result can be given to the user. Finally, if digital real-time feedback should be provided, the designer should choose an appropriate channel following the four-dimensional taxonomy of our model (temporal/spatial distance, modality, data rate).

5. **Leave interactional freedom to the user**
Since the practices of working with pen and paper are highly informal and individual, PPUIs should leave much freedom to the user and impose only minimal constraints on traditional, well-established practices.

In the next chapters, we will integrate the findings presented in the first three chapters of this book, moving on from the analytical and model-centered point of view to a design perspective. We will apply the theoretical model, which was presented in this chapter, and contribute novel interaction techniques and a system framework for paper-based knowledge work.

Chapter 4
CoScribe: A Platform for Paper-based Knowledge Work

In the first chapters of this book, we have set up a theoretical basis for understanding, analyzing and designing Pen-and-Paper User Interfaces. This comprises the affordances of paper (Chapter 1), prior systems (Chapter 2) and a theoretical interaction model (Chapter 3). This fourth chapter and the following chapters serve the purpose of developing a concrete interaction concept that is based on these theoretical underpinnings. The present chapter provides a high-level overview of the entire concept. The following three chapters will then detail on more specific paper-based interaction techniques.

We present CoScribe, an interaction concept and system that supports knowledge workers in dealing with the ever increasing amount of information. Scientists and engineers have early addressed this problem domain, generating such influential visions as those of Vannevar Bush [12] and Ted Nelson [102]. CoScribe addresses this challenge from a paper-centric perspective. It includes novel interaction techniques that tightly integrate the physical world of printed documents and the world of digital documents by offering support for active reading and integration of knowledge from various sources. In addition CoScribe exemplifies how our interaction model, introduced in the previous chapter, transfers to concrete interaction techniques.

CoScribe is largely motivated by results of workplace studies presented in the literature – most important is Sellen and Harper's seminal work [131] – and by a number of field studies that we conducted in our own research. We analyzed how university students use paper and digital media during and after courses, including a questionnaire-based study [145], a contrasting analysis of handwritten and typewritten course notes [143] and an analysis of media use during co-located meetings [144]. Based upon the theoretical interaction model that was presented in the previous chapter, CoScribe takes on an integrated, ecology-centered viewpoint. This means that CoScribe focuses on the interdependencies between (multiple) users, physical and digital artifacts and practices in a given work setting. As a consequence, our concept puts high emphasis on multi-document activities and on collaboration support. We present a set of interaction techniques and visualizations that enable both co-located and remote asynchronous collaboration around paper. We show examples of Pen-and-Paper Interfaces that are built using the core inter-

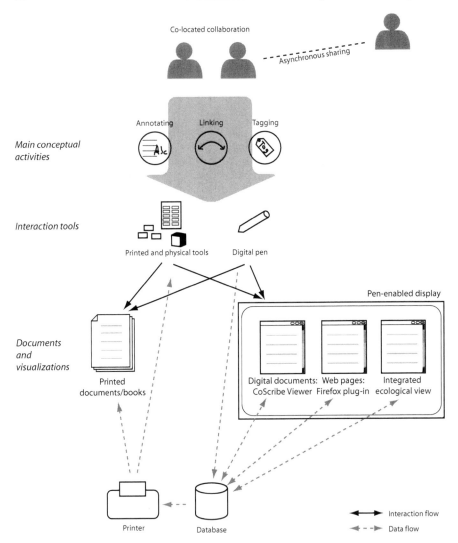

Fig. 4.1 Overview of CoScribe

actions presented in the interaction model. Furthermore, CoScribe simplifies pen-based interaction and bridges the boundaries between paper and the digital world, since the same pen can be used for all activities both on paper and on pen-enabled displays. Finally, our solutions create a richer user experience than previous work by offering a wide set of intuitive tools which are made out of paper.

Figure 4.1 gives an overview of CoScribe, which will be detailed in the following.

4.1 Main Conceptual Activities

Making notes and annotations is a substantial part of reading processes. Opposed to what we intuitively understand by reading – deciphering words and phrases and ultimately meaning – reading documents frequently comprises writing as well. Commenting, underlining and highlighting a document during reading supports better understanding, critical thinking as well as remembering the thoughts the reader had. Adler et al. call this process active reading [2]. As Adler notes, "the physical act of writing, with your own hand, brings words and sentences more sharply before your mind and preserves them better in your memory". Annotations and notes are not only central to reading but also important for efficiently attending meetings or lectures. Psychological research shows that notetaking plays an important role in learning processes and has been proven to be a factor positively related to students' academic achievement [58, 115]. Taking personal notes or annotating handouts stimulates attendees to actively follow the lecture, to consciously select important content, and to rephrase it in their own words. Moreover, notes and annotations have a reminding effect during review after class [59].

In addition to active reading, it is crucial for successful knowledge acquisition to structure the entire problem domain, to abstract and to establish relationships between concepts. The resulting structural knowledge facilitates recall and comprehension and is essential to problem solving [51]. For this purpose, linking and tagging documents are highly relevant activities. These can transform an unsorted and possibly confusing collection of a large number of disparate documents to a unified and well-structured document space.

In practice, annotating, linking and tagging go in hand and complement each other. When reading a document, the knowledge worker would for example make handwritten annotations to highlight important passages and to add some clarifying details. Moreover, he or she would add a reference to another document which covers a particular aspect in more detail. When reading this other document, he or she would make some more annotations before tagging this other document as important and finally going back to the first document.

While there are well-established practices for annotating, referencing and structuring paper documents, these activities are harder to perform with current technology for digital documents. In particular, the transitions between printed and digital documents are not well supported. Notes and annotations made on paper cannot be easily digitized and it is difficult and time-consuming to create references between printed and digital documents. Moreover, while it has become common to tag documents on the Web, these systems cannot be used for tagging content that is available on paper. CoScribe offers cross-media support for annotating, linking and tagging printed and digital documents. These activities enable the knowledge worker to read and understand documents and then to relate and abstract them in order to gain structural knowledge of the problem domain.

Annotating, linking and tagging are generic activities, each of them serving a large number of possible tasks. Therefore CoScribe is not tailored to one specific purpose. Instead, it can be considered as a toolset for document-based knowledge work. The knowledge worker chooses appropriate tools, combines them and uses them in a way that best fits his or her current task. Note that this design is a direct consequence of the ecological perspective. For instance, free-form annotations support tasks such as taking notes during lectures, structuring documents, revising documents, creating summaries or making excerpts. In each of these tasks, the user creates other types and forms of annotations at other document positions. These various practices are possible because annotations can be of any shape and can be made at arbitrary positions within documents. Another example of generic activities are tags. These can be used for such different tasks as conceptually structuring a given domain, defining priorities and to do items or marking up passages that a co-worker should read.

The flexibility of CoScribe is best illustrated by giving examples of settings it can be used in. As our application scenario, we choose learning at universities. The paper-based practices of this application scenario are highly varied and representative for a broad class of collaborative knowledge work settings. Practices include taking notes and making annotations during courses, reviewing own notes and shared notes of other learners, preparing for exams in learning group meetings, excerpting documents, searching and integrating literature for preparing an article or a term paper and even giving presentations by controlling the slide actually being presented using a printout. Despite its wide applicability, CoScribe remains easy to use, as it relies on a small set of simple, but generic interaction strategies, which are inspired from traditional practices of working with paper documents.

Let us consider the following example scenario:

Scenario 1 *Sally is a second-year university student and attends several lectures and seminars. Her work is mainly based on documents, such as physical books, sheets of paper containing her handwritten notes, printouts of web pages, digital PDF articles and digital Web pages. While she often deeply engages with one single document, for example for reading and understanding a complicated scientific article, an important part of her work also consists of integrating contents from various sources to understand a problem domain. She often works on her own, but is also frequently involved in collaborative settings, for example during the sessions of seminars or in meetings with fellow students. She has various workplaces, working at home, in lecture halls, at the library and even in public transport. Hence, the way Sally works with documents and the contexts of this work are highly varied.*

CoScribe supports all these forms of document work. In this and the following chapters, this scenario will be used to illustrate CoScribe's functionality in the light of practical examples.

4.2 Interaction Tools

CoScribe aims at a seamless integration of physical and digital documents. The main interaction tool of CoScribe is a digital Anoto pen (see p. 29 ff. for a presentation

4.3 Synchronized Paper Documents and Digital Visualizations

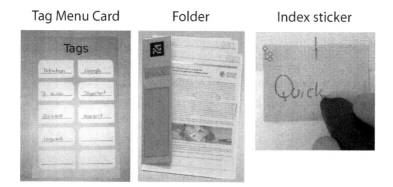

Fig. 4.2 Examples of inexpensive tangible tools that are made of paper

of this technology). The pen is not only used to interact with printed documents and paper tools, but also as a stylus to interact with digital media that are displayed on screens.

Furthermore, CoScribe offers various specialized tangible tools, for instance menu cards, folders and index stickers (see Fig. 4.2). In contrast to the tangible tools used in typical Tangible User Interfaces, our tools are made of paper and moreover do not require additional electronic components besides the digital pen. For these reasons, an ordinary printer suffices to create a large number of tangible tools quickly and at very low cost. CoScribe includes a print toolkit that allows end-users to easily print tools and documents. The toolkit adds the Anoto pattern to the printouts and automatically manages the association between the digital contents and their physical positions on the printouts. This is necessary for mapping pen interactions on paper to the underlying digital contents. All interaction techniques of CoScribe make consistent use of the core interactions presented in the previous chapter.

4.3 Synchronized Paper Documents and Digital Visualizations

Users can print digital documents onto real paper and interact with them using an digital pen. Due to the Anoto technology, it is possible to simultaneously work with several sheets of paper and to use the same pen on all of them without additional effort. This provides for a very natural working style.

While paper provides for a flexible use of documents, there might be situations in which the user prefers working with a digital version of a document. For example full text search as well as audio and video contents are better supported by digital documents. Moreover, digital visualizations allow the user to access shared annotations of other users in real-time. For these reasons, CoScribe does not only support printed documents. The user can access a digital version of any document in a software viewer. This version includes own and shared annotations, hyperlinks and tags. To support both autonomous mobile use and real-time interaction with the system, CoScribe offers two ways of synchronizing pen data with the digital system:

1. *Batch synchronization:* In order to retain the mobility which is inherent to paper, the digital pen can be used autonomously. During writing on paper, data is buffered on the pen. At regular intervals, the user synchronizes the pen with a computer. All data temporarily buffered on the pen is then transferred to the digital system. Batch synchronization supports particularly well writing tasks, which do not require immediate system feedback. However, no real-time interaction with the digital system is possible.
2. *Real-time streaming:* If a computer is nearby, pen data can be sent in real-time to the digital system over a wireless Bluetooth channel. This has the advantage that all data can be immediately processed. For instance this is important for providing real-time feedback when the pen is used as a stylus on a pen-enabled display. However, the mobility of paper is restricted to an area within the range of the wireless connection. With the current Bluetooth specification, this is up to distances of about 100 metres. In the Bluetooth setup, up to 8 pens can simultaneously connect to one single client computer. This provides for co-located use. If a larger number of pens is to be used, several client computers can be used in the same place, which synchronize their data via the central database. [1]

Once the pen data is transferred to the client computer, it is automatically handled by a stroke processing module. This module checks whether the stroke originates from paper or from the pen-enabled display and maps it to the associated digital contents or interface elements. It then interprets the interaction and, if necessary, executes a command. If the stroke is not a command but an inking interaction, individual strokes are clustered into annotations.

Software Viewers Changes made on a printed document are automatically included in its digital version and made available in a software viewer. CoScribe includes several viewers for different types of documents. Currently supported document types are PDF and PowerPoint documents, Web pages and physical books. For PDF and PowerPoint documents, CoScribe comprises an own document viewer (Fig. 4.3). While we opted for a proprietary solution for our prototype, enabling more flexibility during the iterative design process, future versions of CoScribe could use plug-ins for standard software, such as Adobe Acrobat and Microsoft PowerPoint. Web pages are available in Mozilla Firefox; a CoScribe Firefox plug-in enables pen-based input on Web pages and displays additional information, such as hyperlinks created by the user. A third viewer, the ecological view, integrates all documents and displays them in a graph-based visualization (see Section 6.4). The viewers can be used on standard computers. If the computer features an Anoto-enabled display, they can be controlled with the same pen as used on paper. The

---

[1] Note that currently only one Anoto Pen, Nokia SU-1B, allows for live switching between batch and streaming mode. Other pens can be used only in one of the two modes. This is a technical limitation of current pen models, not of our concept.

4.3 Synchronized Paper Documents and Digital Visualizations

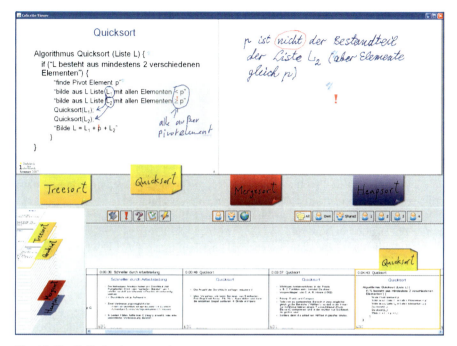

Fig. 4.3 The CoScribe document viewer

same pen gestures as on paper create and modify annotations, links and tags directly on the digital documents.

In contrast to changes made on the printed representation which are automatically transferred to the digital representation, changes made within the viewer are not automatically reflected on paper. If a corresponding printed representation is needed, one can use the printout module to print an updated version of the document – the paper-digital document cycle is complete.

In summary, the user can choose between working with a printed representation of a document and with its digital representation, depending on his or her preferences and the context. Physical and digital versions can also be used simultaneously.

Transformations between Paper and Digital Views Working with both printed and digital versions of the same document requires easy access to the corresponding versions. The following example scenario shows when this can be relevant:

Scenario 2 (Interwoven Use of Printed and Digital Representations) *Sally hopes finding further helpful annotations of other students. Therefore she browses through the digital version of the lecture slides. After some time, she comes across a very helpful annotation. She decides to add this annotation directly to her printed script. She therefore searches the sheet of paper containing the slide and writes the annotation on it.*

In CoScribe, all printed documents feature a printed button that allows for quickly accessing a digital version of a printed page. By tapping with the pen on that button (Fig. 4.4 upper right), the digital counterpart is displayed in the CoScribe viewer. A second button (upper right) allows for accessing the ecological view, which provides an overview of the entire document space. The thumbnail of this particular document page is then centered and highlighted in this view.

For the reverse direction – accessing a printed representation of a digital document – the user can print a new copy of the document. However, it is impractical, expensive and harmful to the environment to re-print the information at a frequent basis. Another way of transformation is to access an already existing printed representation. This implies that the user must find this physical copy, which can be hard if the user disposes of a large number of printed representations.

A number of approaches have been presented that automatically track the location of physical objects in the space, e.g. by using RFID tags [22]. This knowledge is used to indicate the location, for example by lighting up an LED which is located at that position. In order to do so, the system must track the location of individual sheets of paper. Since the technical setup for tracking the location of physical objects strongly restricts mobile use, CoScribe is based on another approach. This does not indicate the concrete physical location but provides information about properties of the representation, such as its visual appearance. The user can then utilize this information to easily find the physical representation. CoScribe uses the following properties:

- The *visual appearance* of the document (the thumbnail or document page view visually corresponds to the printed page). This is particularly helpful if the layout of different documents and pages is diverse. In this case, the size of the margins, the number of columns and the repartition of headings, paragraphs and illustrations can provide a good visual cue for finding a given document or page.
- If the document is contained within a *folder* (see p. 136), the system indicates this folder. As the number of folders is typically much smaller than the number of documents, this simplifies finding the document page.
- Digital Paper Bookmarks are adhesive sticker which the user can attach to individual document pages in order to index them. These stickers are well visible because they jug out of the document. The *arrangement of bookmarks* on a document as well as their colors and labels provide good indicators for identifying a

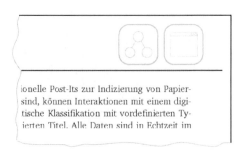

Fig. 4.4 Buttons for accessing the digital equivalent of the physical page in the document viewer (right) and in the ecological view (left) are printed in the top right corner of each document page.

specific document at a glance. More details on Digital Paper Bookmarks and the corresponding digital representations are given in Section 7.1.

4.4 Collaboration

A further aspect which is essential for gaining new knowledge is collaboration. By the exchange with other people, knowledge workers can gain new insights and perspectives, can critically examine their own understanding and can co-construct a shared understanding with others. While the functionality of CoScribe is helpful for individual users, its strength is in supporting collaboration between multiple users. On the one hand, users can collaborate in a co-located setting, for example in meetings:

Scenario 3 (Learning Group Meeting) *During the weeks before the final exam of a lecture, Sally regularly meets with two fellow students in an open space at the library. Together they review the lecture handout and their annotations and discuss unclear topics. If necessary, they collect further information from textbooks and Web pages and link these to their lecture handouts. At the end of each meeting, they collaboratively create a summary of the topics.*

In this scenario, several persons use the system at the same time in the same place. They interact with printed documents on paper and with digital documents on one or more pen-enabled displays (Fig. 4.5). Each user has his or her own pen. In this setting, CoScribe is designed for a rather small number of users (about two up to six users). In a small group, users can physically share their documents.

Fig. 4.5 Co-located collaboration

Even though traditional paper is well suited for many types of co-located collaboration, it obstructs remote collaboration, as it is more difficult to share physical

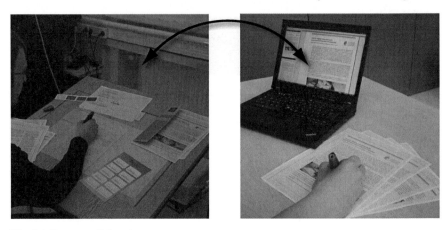

Fig. 4.6 Remote collaboration

contents over distance than digital information. CoScribe also supports this mode of collaboration, for example to continue collaboration after a co-located meeting:

Scenario 4 (Accessing Documents at Home) *Dan is ill and could not attend today's learning group meeting. He uses the CoScribe viewer to access the summary document created by the other team members.*

Different users can access digital versions of the documents (and print these, if desired). The documents include annotations, links and tags that were made by other users. In a remote setting, CoScribe supports a very large number of users, ranging up to several hundreds of users in a large lecture. While it provides some support for remote synchronous settings, asynchronous forms of remote collaboration are more adequate for shared annotations that heavily rely on paper. This is due to the fact that contents on paper cannot be updated in real-time by the digital system. In contrast, it is not problematic to re-print updated versions once in a while.

In order to provide efficient remote access to shared contents of many users, several collaborative visualizations for documents, annotations, hyperlinks and tags have been developed which integrate user-generated content of several users: First, the user can print an updated version of the document which includes the annotations, links and tags of one or more other users. Second, user-generated content of other users is automatically integrated into the digital representation of a document, which is displayed in a software viewer. Third, an ecology-centered visualization integrates all documents and user-generated content from all users. It provides overview of and structured access to the entire collection of documents. This is particularly supportive if a considerable number of users collaborate on a large number of documents, a setting where it might be difficult to find a given document or a given passage of a document at a subsequent point in time (see Section 6.4). Finally, further more specialized visualizations focus on specific aspects. For example, one view integrates the bookmarks made by all users on a particular document in order to support users in comparing the structuring created individually by each user (see Section 7.1).

4.5 Implementation

CoScribe was implemented as a research prototype. It is realized as a client/server system that includes both software components and specific hardware components. Figure 4.7 depicts a deployment diagram that gives an overview of the components. Most components of CoScribe are implemented in Java; the modules for pen connection and handwriting recognition are implemented in C#, the Firefox plug-in in JavaScript. We use an OKI C5900 color laser printer for printing documents and paper tools. Our prototype has successfully been tested to work with the following pen models: Anoto DP-201, Logitech io2 and Nokia SU-1B.

As shown in the deployment diagram, most components of CoScribe are executed on a local client. However, a central repository on a server is accessed by all local installations of CoScribe to exchange shared data. We use a graph-based database system, which was developed at our institution. This system automatically maintains a local mirror of the central repository (containing all information the

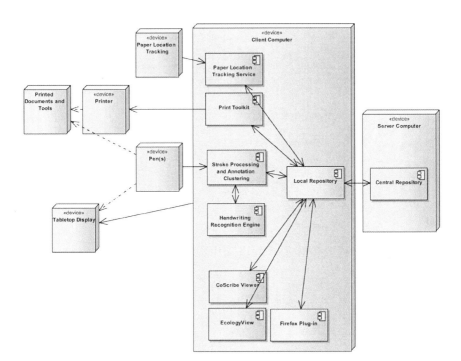

Fig. 4.7 Overview on the software and hardware components of CoScribe

specific local user has the right to access). This provides for using CoScribe in a single-user configuration without a central server and for coping with temporary unavailability of the internet connection (e.g. during mobile use).

The following three chapters will provide a more detailed discussion of CoScribe's paper-based interaction techniques for annotating, linking and tagging documents. Where appropriate, we will also describe the technical components in more detail.

Chapter 5
Collaborative Cross-media Annotation of Documents

The discussion of the affordances of paper (Section 1.1) has shown that one of the main reasons for the longevity of paper is that paper documents can be easily annotated. Annotations serve many different purposes. The following scenarios illustrate some of them in our application domain:

Scenario 5 (Mobile annotation) *Before a lecture, Sally prints the lecture script, which is distributed online by the instructor. During the lecture, she then makes annotations on this script using her digital pen. She does not need a computer, but only the digital pen and the printout. As she lives at some distance from the university, she spends a considerable amount of time in public transport. On the way home, she typically uses this time to review the contents presented during the day. Using the digital pen, she marks up passages of the lecture script which are particularly important or which she has difficulties understanding.*

Scenario 6 (Active reading) *Sally has the task to read an article before the next lecture. It is a challenging text, which has a complex structure and includes many foreign words. Sally therefore works intensely on the text. During reading, she makes annotations: She underlines key words and marks up important passages. She looks up the signification of foreign words and annotates them with the corresponding English terms. Moreover, she adds clarifying details and notes her own thoughts and ideas concerning the text.*

Scenario 7 (Reviewing annotations) *At home, she puts the pen into a reading device, which automatically transfers all pen data to the CoScribe client which runs on Sally's computer. She then consults a textbook and completes the notes she has made during the course. The pen is now used in live-mode, i.e. all data is available in CoScribe in real time. As she has not well understood one particular slide, she opens the CoScribe document viewer and reads the annotations that other students have made on this slide.*

In this chapter, we present a set of interaction techniques and visualizations for cross-media annotation of documents. We contribute to the notable body of existing research in two respects: First, we bring the layout of the annotation interface into

Table 5.1 Desiderata and our approaches concerning annotation of printed and digital documents

Desideratum	Approach	Section	Concept	Function	Innovation w.r.t. related work
Flexible free-form annotations	Unrestricted PPUI; printing documents with annotations	5.1	•	•	
User-adaptable PPUI	Various print layouts	5.1	•	•	
	Free arrangements of several pages	5.1	•	•	•
Paper-based sharing and classification of annotations	Tagging with Buttons	5.2	•	•	
Integrated access to shared annotations	Integrated visualization using dynamic shrinking/expanding and repositioning of annotations	5.3	•		•
	Integrated multi-user visualization	5.3		•	•
	Integrated multi-user printouts	5.3		•	•
Access to text of handwritten annotations	Evaluation and use of handwriting recognition	5.4.3	•	•	

the focus of attention. This aspect was not considered in previous systems, yet has a significant impact on the annotations made. Second, we address asynchronous sharing of annotations and present a paper-based sharing method as well as a novel collaborative visualization that integrates handwritten annotations of several users. We conclude by presenting and discussing the results of three evaluation studies. Table 5.1 provides an overview on the challenges which we address in this chapter.

5.1 An Adaptable Printed User Interface for Annotations

Our design was guided by the goal of providing for a smooth transition from traditional paper annotations to a computer-supported collaborative annotation process, permitting each user to maintain his or her personal annotation style. Handwritten annotations present a large variety in form and content [91]. Figure 5.1 depicts several examples of annotations made on course materials. The examples show that while many annotations are quite condensed, containing only some keywords, annotations can also become very extensive, covering an entire page in small hand-

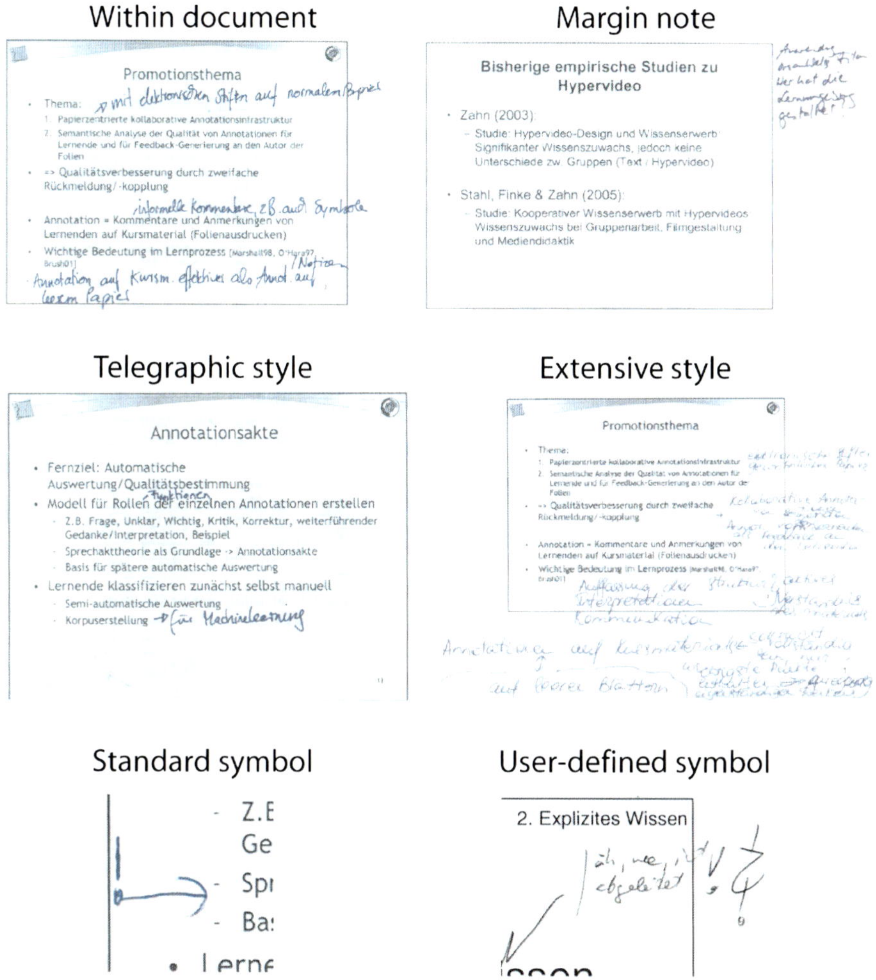

5.1 An Adaptable Printed User Interface for Annotations

Fig. 5.1 Examples of handwritten annotations

writing. Existing paper-based annotation interfaces do not address this issue and provide to all users the same annotation space. Initial field analyses undertaken in the early design phase of CoScribe showed that depending on their personal annotation style, users prefer different print layouts. For instance, our observations showed that additional empty areas on lecture handouts seemed necessary to most students for making extensive annotations. Most right-handers prefer such empty notetaking areas being located to the right of the actual document page, while left-handers prefer them to the left. While in some contexts, the user makes many annotations on the document (e.g. during a lecture), other contexts (e.g. excerpting on a separate sheet of paper) do not require making many annotations and therefore, the user should be

Fig. 5.2 Annotating a printed document using a digital pen

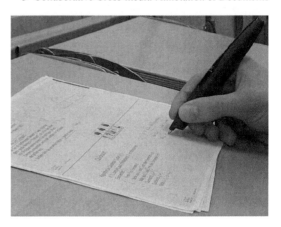

enabled to print pages without additional notetaking ares or even in a down-scaled manner.

Our interface offers similar basic functionality as paper-based annotation systems from related work. Users can print digital documents on paper and annotate them with the digital pen (Fig. 5.2). At any time, the user can add, modify or delete annotations. An annotation is deleted by performing a cross-out gesture on the annotation. After the pen data has been synchronized, the annotation is then removed from the digital viewer and not printed in subsequent printouts. A new printout of the document can optionally include the already existing annotations. Improving over previous work, the print module can visually communicate meta-information about annotations which was not visible on the original document: The color of the annotation encodes its category. Moreover, the older an annotation is, the lighter it is printed, inspired by how an old document optically bleaches out. Finally, collaborative annotations that were read by a very large number of users are printed with thicker pen traces.

In addition to this basic annotation functionality, the user can adapt the Pen-and-Paper User Interface of CoScribe to fit his or her personal preferences and the current context. This is supported both at the levels of individual pages and of the ensemble of pages used:

At the level of individual pages, users can choose among various layouts of printed document pages. For instance, each printed page can contain one page of the document or alternatively, several document pages can be printed on one single sheet of paper in a down-scaled layout. Optionally, empty noteteking areas can be included. The position of other interface elements, such as button toolbars, can be equally modified. Figure 5.3 gives several example layouts. Our technical realization relies on the concept of logical and physical pages. A logical page corresponds to one page of a physical document. Several logical pages can be combined on one physical sheet of paper. Thereby each logical page can be freely scaled and positioned on the physical page (see Fig. 5.4). Moreover, within each logical page, the actual page of the digital document can be freely positioned and scaled. This pro-

5.1 An Adaptable Printed User Interface for Annotations

Fig. 5.3 Example layouts of printouts

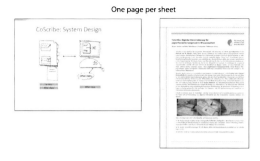

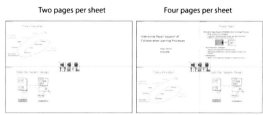

Fig. 5.4 Example of physical and logical pages

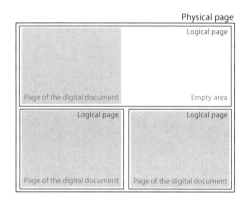

vides for reserving empty space for taking notes beside, above, below or around the document page. Annotations are stored in a generic representation, independently of the currently chosen print layout.

In contrast to other applications, which usually model handwritten annotations as being located on one single page, our model supports annotations that span multiple pages. This is particularly important because one physical page can contain multiple logical pages and users frequently write over the boundaries of individual logical pages. It would not be acceptable to cut these annotations into two or more separate fragments. Instead, our model keeps these annotations intact and treats them as one single annotation. The x and y coordinates of the individual samples of the annotations are normalized to be in a $[0..1]$ range, where the coordinate $(0,0)$ denotes the upper left and $(1,1)$ the lower right corner of the bounding box of the annotation. This makes the modeling independent of the actual position of the annotation on a page. For each logical page the annotation is located on, it contains some context information that indicates which portion of the annotation is visible at what area of this page.

Adapting the print layout implicates the drawback that the user has to make a decision before printing the document. Therefore, in addition to adapting the layout of single pages before printing them, users can dynamically modify the layout of several printed pages. If the space available on a document page (and on the optional notetaking area) is not sufficient for making annotations, users can dynamically add one or more empty paper sheets to this document page by associating them with a line gesture (see Section 6.2 below). This corresponds to the GUI interactions of enlarging a window to have more space available or to scrolling in a document view to display an empty area.

5.2 Paper-based Sharing of Annotations

As discussed in the previous chapters, collaboration is an important element of document-based knowledge work. While paper documents are well-suited for many types of co-located collaboration, they constrain remote collaboration in comparison to digital documents. This section focuses on remote sharing of annotations and discusses means that allow the user to share annotations directly when writing them on paper.

Scenario 8 (Private and Public Annotations) *It is absolutely acceptable for Sally to share her annotations of the lecture script with the other members of her learning group. However, from time to time she makes off-topic notes, which she prefers to remain personal. For example, she makes an appointment with a fellow student and notes his phone number. Moreover, she takes a note reminding her what she wants to buy after the lecture. She marks these notes as private. They are not shared with other persons.*

An interaction technique for sharing notes should be seamlessly integrated with annotating and be quick and reliable. Related research [74, 75] discusses several means for classifying annotations:

One method is spatial differentiation. This consists of reserving different areas on the printout for annotations of different visibility. The annotation has automatically the visibility of the area it is written in. For instance, all annotations written in the "share" area are automatically shared with collaborators. Spatial differentiation is intuitive and provides clear visual feedback on which annotations are shared. However, it is not possible to make an annotation directly within its context of the printed document, e.g. highlighting a specific word. Moreover only a very small number of sharing levels can be supported because each additional level requires an additional empty area which occupies valuable space on the printed document.

A second method consists of using different pens for different visibilities. Each pen is associated with one visibility. While for instance all annotations written with the green pen are automatically shared, annotations written with the red pen remain

5.2 Paper-based Sharing of Annotations

private. Pen switching is also intuitive and provides clear visual feedback. However, the user has to buy several digital pens. Moreover, research shows that people tend to use one single pen rather than switching between many different pens [91]. Switching between different pen modes using one single pen, e.g. by pressing a barrel button on the pen, might alleviate this issue but only allows for a small number of visibilities.

A third method relies on buttons that are printed on the printout. Each button represents one visibility level. By tapping on that button before writing the annotation, the respective visibility is set. The advantage is that a large number of visibilites can be supported using only one single pen. Moreover, performing a simple pen tap is quick and can be easily included into the annotation process. However, no visual feedback on the visibility level is available on the printed document.

A fourth method consists of classifying notes by performing specific pen gestures. While this requires memorizing specific gestures, a potentially large number of visibility levels can be supported. The gestures also visually communicate which visibility has been defined. However, as gesture recognition involves uncertainty, the user must be provided with instant feedback on whether a gesture has been correctly recognized. This requires specific digital pens that can perform gesture recognition and provide visual feedback directly on the pen, even in mobile conditions without another computer around. Currently only Livescribe pens have these capabilities. Moreover, the system has to distinguish gestures from ordinary handwritings or drawings. For this purpose, related research suggests to use additional hardware like a foot pedal or a second pen [75].

All of these methods can be combined with CoScribe and are compatible with the collaborative visualizations presented in this chapter. In our current implementation, we opted for the button-based method. It would have been to costly to equip a large number of participants of our field studies with multiple pens each to allow for pen-based differentiation. Moreover, at the time of our evaluations, Livescribe pens were not yet available on the market[1]; so in the case of gesture-based differentiation no immediate feedback on the result or failure of recognition could have been given.

A toolbar containing several printed "buttons" is printed in the center region of each paper sheet (see Fig. 5.5). A visibility is associated to an individual annotation by tapping with the pen on the corresponding button before writing the annotation. Moreover, a visibility level can be set or modified later on by making two consecutive pen taps on the button and on the annotation. While no graphical feedback on the tagging is provided on the printed slide unless the user makes additional markings, the visibility level is visualized with specific colors in the CoScribe viewer and on subsequent printouts. The viewer contains similar buttons as the printouts for defining or modifying visibilities.

A correct interpretation is guaranteed, as determining the pen position on a paper button does not imply uncertainty. Most pens provide graphical feedback to the user at the moment the button area is tapped on (Fig. 5.6). Hence, the user can be sure the classification has been correctly recognized. A problem of most current Anoto pens

---

[1] Even current Livescribe pens do only partially support our concept, since they do not allow for live streaming of pen data.

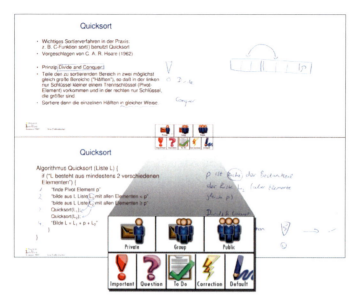

Fig. 5.5 A button toolbar is printed on each page. These buttons provide for defining the visibility level of annotations (upper buttons) and for tagging them with semantic types (lower buttons).

is that they cannot provide feedback on the current mode. Therefore the selection of a visibility level is not permanent in our design. The visibility only applies to the annotation created immediately afterwards, then the system returns to the default classification state. Thus the user does not have to remember a current system state. If the user desires absolute certainty to be in the default mode, he or she can tap on a 'default' button. While a completely modeless design would be preferable, this "semi-modal" design accounts for the two stages of selecting an instrument and an operand and nevertheless copes with the absence of feedback. Novel generations of digital pens that include a display will solve this problem. They will make the use of button-based classification more comfortable and even allow for live recognition of gestures.

Applying visibilities to entire annotations instead of unstructured sets of pen strokes requires clustering pen strokes into annotations. For this reason, CoScribe uses a clustering algorithm for handwritten input that relies on temporal and spatial information. Possible errors can be manually corrected by the users. The software viewer therefore provides two functions for splitting annotations and merging pen strokes into one annotation.

Our implementation offers three levels of visibility with which the user can classify individual annotations:

- *Private visibility*: The content is not shared with other users.
- *Group visibility*: Each user can set up groups with several other users. In different contexts (e.g. a specific lecture, a meeting series etc.), the user can be member of

5.3 Visualization of Shared Annotations

Fig. 5.6 A LED (within the red circle) lights up when the pen taps on a printed button and provides visual feedback on the selection

different groups. Content with group visibility is shared with the members of all groups of this user that apply to the given context.
- *Public visibility*: The content is visible to all users in this given context (e.g. all students attending this lecture, all participants of the meeting series etc.).

Moreover, defining visibility is optional, allowing the user to maintain a natural annotation style. If no visibility is chosen, the default level defined by the user is set. This reduces extraneous cognitive load during the annotation process. Results from semi-structured interviews that we conducted during our evaluation studies indicate that a very appropriate default level is the group level. With respect to privacy, it is typically not considered critical to share annotations with other members of the own group, as these are personally selected by the user. With this default level, only a small number of annotations that contain private information or that shall be visible to all users must be explicitly classified with a visibility level.

5.3 Visualization of Shared Annotations

Accessing annotations that were shared by other users can support understanding of documents:

Scenario 9 (Thoughts of other Persons) *Sally has not well understood a particular issue. Reading the annotations of other persons helps her in getting the point. In another course, a controversial topic was discussed. She is interested in what the other students think of this point and reads their annotations.*

Scenario 10 (Catching up on a Lecture) *Sally was ill for some days and missed several lectures. She therefore carefully reads the annotations that her learning partners have made during the lectures.*

Scenario 11 (Collaborative Notetaking) *Sally attends a course in which the handout is not very detailed. For this reason, the students take very extensive notes and sometimes there is not sufficient time to note everything which is important. Hence, Sally and two other students have agreed to jointly take notes. Each person is responsible for one particular aspect and notes everything which is related to this aspect. After the course, each of them completes his or her own notes with the notes of the two other students. This leads to more comprehensive notes.*

In this section we discuss how to effectively visualize shared handwritten annotations that multiple users have made on their personal copies of the same document. An aggregate view that integrates handwritten annotation would quickly become cluttered or even illegible when displaying a large number of possibly overlapping annotations of different users directly on the document. We present a digital and a paper-based visualization which integrate shared handwritten annotations more effectively.

The CoScribe viewer provides digital access to both own and shared annotations. It includes a *single-user view* for each member of the user's learning group. This separates the annotations of different users into different views. While each of these views in itself becomes easier to read, this implies the need to manually switch between views of different users. This becomes particularly cumbersome in larger communities. For this reason, CoScribe also features a *multi-user view* which combine both one's own and shared annotations in an integrated visualization. We examined three different approaches. First, overlapping annotations could be moved to the margin of the documents. This implies the drawback of separating annotations from their context. Annotations that visually refer to the context (e.g. underlinings, arrows, captions for printed elements) can become illegible. As a second solution, the white spaces within a document page could be stretched to provide enough space for all annotations. However, this could result in very large document pages in cases where are many annotations. Moreover, multiple annotations that refer to the same element might loose their context.

We opted for a third technique which consists of varying the size of individual annotations: collapsing annotations that are currently not relevant and expanding annotations that are in the user's focus. In the CoScribe viewer, one's own annotations are visualized as they are written on paper (Fig. 5.7 (1)), whereas shared comments of other users are displayed in a condensed form. Instead of the annotation itself, a small icon is visualized at the position of the annotation (Fig. 5.7 (2)). This icon corresponds to the annotation category and varies in size according to the size of the annotation. When hovering with the mouse over the icon or tapping with the pen on it, the annotation is expanded and displayed within its context and in its original size (Fig. 5.7 (3), annotation with grey background). The user can copy shared annotations that are considered especially relevant to his or her own script. They are then displayed in their decompressed form like one's own annotations. Moreover,

5.3 Visualization of Shared Annotations

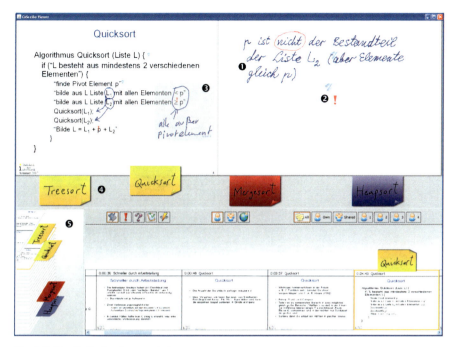

Fig. 5.7 The CoScribe document viewer

the symbols of annotations that are not considered relevant can be removed from the view.

This has the advantage that the entire document remains readable and is displayed at a reasonable size. However, not all annotations are visible at the same time. For quickly gaining an overview of all annotations, CoScribe includes a preview function, which displays all shared annotations in small size. Moreover, a view of all annotations made by a particular user is displayed when pressing a specific key while hovering over or tapping on an annotation of that user. This way, a single-user view displaying only the annotations of one specific user can be easily accessed.

The multi-user and single-user views complement each other. While the multi-user view serves well for getting an overview of all annotations on a page, the single-user view is better suited when specifically focusing on the annotations of one specific user, for instance those of the instructor or those made by a student known for making helpful annotations.

CoScribe offers representations for printed documents that are equivalent to the digital single-user and multi-user views. Users can print documents including the annotations of one single user. This corresponds to the single-user view. In contrast, printing the annotations of multiple users poses the same challenges as on the screen because annotations of different users might overlap. However, in contrast to screens, dynamic scaling of annotations is not possible. For this reason, the print module generates a layout in which the document page is slightly scaled down.

Fig. 5.8 A printout including annotations of two users

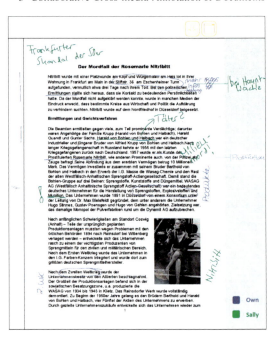

Of each set of overlapping annotations, only one annotation is printed at its correct context position, while all other annotations are moved to the margins. Their context position is marked by a connecting line (see Fig. 5.8).

An important issue of views that integrate annotations from all users is their scalability to a large number of annotations and users. An evaluation with annotations made by students in real lectures (which is presented in more detail in Section 5.4.1) supports the assumption that in a lecture scenario, the views scale well to a larger user community. The reason is that even in very large audiences, the average number of shared annotations per document page remains rather small. This will be demonstrated by the following example calculation: Scalability clearly does not depend on one's own annotations. Similarly, annotations which are made by members of the user's group do not have an influence, since the average number of members in a group is not affected by the size of the entire audience. Hence, scalability only depends on public annotations. Our experiences show that in a lecture setting, only a very small fraction of annotations is published to the whole community. In our case this were 1.6 % of all annotations. In our lecture evaluation, each student made an average of 0.59 annotations per slide. Assuming four members in a learning group, an average of 1.7 additional annotations are shared by these members. Public annotations of the entire audience average out at 0.9 annotations for 100 participants and at 4.7 annotations for 500 participants. Hence, the total number of annotations keeps quite small.

There are however some slides which are heavily annotated. In our evaluation, the most frequently annotated slide contains an average of 5.2 annotations per user.

This "worst case" results in an average of 15.2 annotations shared by the own group and an additional 8 annotations for 100 participants (or 41 annotations for 500 participants). These high numbers apply only to a very small number of slides (only the top 10 % of slides have more than an average of 2.8 annotations per user). In order to cope in these situations with a too large number of public annotations, these annotations could be automatically filtered. For instance, while personal and shared group annotations are visualized as discussed above, only these public annotations that have been classified as relevant by members of the author's group are displayed to the entire audience. Another approach could consist of automatically summarizing contents of annotations. This however requires quite reliable handwriting recognition.

5.4 Evaluation and Discussion

We conclude this chapter with the results of three evaluation studies of CoScribe. The studies aim at evaluating CoScribe's annotation functionality. In two user studies, we evaluated the usability of CoScribe for within-lecture annotations and post-lecture review. Our main goal was to examine whether the interaction techniques are efficient, reliable, easy to learn and easy to use. A second main goal of the user studies was to assess user satisfaction in order to examine if the novel interaction techniques and visualizations are accepted by the users. In this respect, subjective feedback was an important instrument. Further, we aimed at gaining first user experiences and feedback on potential for improvements. In a third study, we focused on handwriting recognition. We analyzed the recognition accuracy of handwritten annotations that were made on lecture slides. Based on these results, we present an approach that considerably increases the recognition accuracy for domain-specific terms.

5.4.1 Study I: Field Study of Lecture Annotation

A first user study examined the use of CoScribe for annotating documents in the field, as a tool for annotating lecture slides during regular computer science lectures. Our goal was to assess the ease of learning and the ease of use of the printed user interface and of the interaction techniques for making annotations and for classifying them with visibilities. A further question was whether the techniques can be easily integrated into the ecology of lecture notetaking, which is characterized by a high degree of intrinsic cognitive load. We moreover analyzed the annotations made by the participants in order to assess types and frequencies of annotations made during the lectures.

Method

We selected three computer science courses of different years of study: a second year introductory lecture on sorting algorithms, a third year lecture on computational engineering and a lecture for graduate students on ubiquitous computing. A total of 29 students (5 female, 24 male) were recruited among the attendees of these lectures. Their semester of study averaged out at 5.6 ($SD = 3.1$). Participation was voluntary and no compensation was given. To avoid bias, none of the students was personally known to us nor did they attend one of our courses.

Before the lecture, each participant received an Anoto pen and a digital paper printout containing the slides of the current lecture. Each A4 sheet contained two slides on the left-hand side and empty areas for taking longer notes to the right. The user was trained for three minutes on how to make annotations and to classify them using the digital pen. The task during the lecture was to make annotations on the printout the same way the participants usually do. In addition, we asked them to semantically classify and/or to share annotations using the paper buttons when it seemed appropriate to them. All these activities were digitally captured by the CoScribe system for subsequent analysis.

After the lecture, the participants filled in a standardized questionnaire for quantitative feedback on the paper-based user interface. The questionnaire contained 25 closed and open questions related to the printed interface, to the digital pen, to the lecture and to personal information. Finally, in each of the three sessions, we conducted a semi-structured group interview with three to six participants. The goal was to gather additional qualitative insights into benefits and shortcomings of the current design and to brainstorm about which further functionality would be helpful. These interviews were videotaped and varied in length from 35 to 70 minutes.

In the statistical analysis of the questionnaire data, we investigated correlations between Likert-scale items which were five-point scaled and performed χ^2-tests and t-tests to identify significant group differences. All these tests were based on a level of significance of 95 %. Moreover, we analyzed the annotations that were made during the lecture. This comprised a statistical analysis of frequencies, positions and types of annotations.

Results and Discussion

Document Annotation Participants considered digital pen and paper for annotating lecture slides to be easy to use. All users reported that annotating printed lecture slides with the digital pen worked reliably and as they had expected. In the questionnaire, the participants judged the use of CoScribe about as distracting as traditional pen and paper ($M = 2.7$ on a scale ranging from 1=more distracting to 5=less distracting, $SD = .7$, $N = 29$), but significantly less distracting than using a laptop ($M = 4.5$, $SD = 1.0$, $N = 29$) ($T = 9.0$, $df = 26$, $p < .001$) (see Fig. 5.9). This indicates that the novel techniques can be well integrated into the existing ecology.

5.4 Evaluation and Discussion

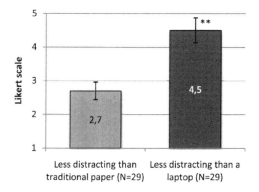

Fig. 5.9 Comparing annotating with CoScribe with traditional pen and paper and a laptop (error lines indicate the 95 % confidence intervals of the means)

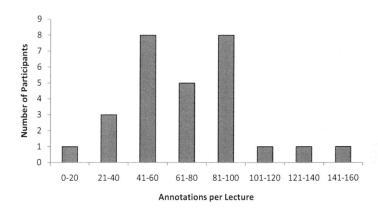

Fig. 5.10 Histogram of the total number of annots. made per participant during a 90 min. lecture

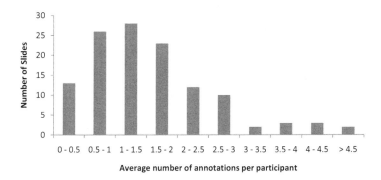

Fig. 5.11 Histogram of the average number of annotations each participant made on a single slide

Although the users have had only a few minutes for familiarizing with the system and used it during one of their normal lectures, they made a considerable amount of annotations. We collected a total of 1983 handwritten annotations during three lectures. In average, each user made 68 annotations ($SD = 29$, $N = 29$) during a 90 minutes lecture. Figure 5.10 shows a histogram of the number of annotations that each user made during one lecture. A high variability between participants was encountered, ranging from only 7 up to 141 annotations per participant. This is in-line with findings reported in the literature [109, 92]. In contrast, we found no significant differences between the different lectures. On average each user annotated 63 % of all available slides, making 2.6 annotations on each of these slides ($SD = 1.7$). Figure 5.11 shows a histogram of how many annotations were in average made on a slide by a single user. No relation between the time within the lecture and the number of annotations was found. All slides were annotated by at least one participant. The ten most frequently annotated slides contained an average of 41 annotations ($SD = 5.8$), while only 2 of the 122 slides contained one single annotation. This shows that the participants were active throughout the entire lecture and that no temporal "hot-spots" of very high or very low activity existed.

Advantages mentioned in comparison to electronic systems without paper support consisted of the possibility to easily make annotations during class and to work with formulas. For instance one student reported in the interview an experience with a Web site designed for group work of mathematics students, which was not accepted and not used by the students since it was by far too complicated to enter formulae with the keyboard. Advantages with respect to pure paper environments were better orientation in the digital document, quicker access in later semesters without the need to keep a physical folder in range and better support for group work. Several users reported that a drawback of the system is that the Anoto pen has only one fix color. Instead, they would prefer a pen that can switch between various colors. This is an area for improvement of Anoto pens. Another problem was that many users felt irritated because the pen vibrated sometimes when it was used for writing over printed text, which is a technical problem occurring from time to time with currently available Anoto pens.

With respect to the annotation of printed documents, CoScribe covers an aspect which is highly important for efficient use and which has not been addressed by previous work: the layout of the printed documents. The participants had different preferences concerning the layout of printed slides, which underscores the importance of flexible layouts. Roughly 75 % of the participants reported in the interviews to prefer only two slides per paper sheet, as this leaves free areas for annotations. The remaining participants preferred four or more slides per page. All left-handers wished having the free annotation areas to the left of the printed slides, while right-handers preferred them to the right. Positioning the free areas below instead of besides the slides was judged less appropriate, since the items on the slides are typically organized in vertical order and it would therefore be more complicated to relate notes to individual items.

Paper-based Sharing of Annotations We assessed the percentage of annotations that were tagged with visibilities. Tagging with visibilities was performed only for a small percentage of annotations. An average of 2.4 % of all annotations was classified as private. An average of 1.6 % was classified as public. Despite these low scores, we do not conclude that privacy mechanisms are not necessary. In all interviews, it became obvious that the users require a functionality for defining the visibility of annotations. Rather, the low scores reflect that the default setting of group visibility is appropriate for most annotations made during a lecture.

In the interviews, there was a wide range of responses to the functionality for classifying annotations. While nearly all participants agreed that this is an important feature, that tapping on a button is quick and easy and does not disrupt the main task of annotating, they disagreed about whether the system feedback is sufficient. Many users reported to feel unsure whether a printed button has been correctly activated when tapping on it with the pen. While the pen confirms the pen tap on a button by briefly lighting up a LED, it provides no feedback on the currently activated classification mode. This was not possible with the Anoto pens available at the time of the evaluation. Novel pens with integrated feedback capabilities could make the classification with paper buttons more reliable.

5.4.2 Study II: Laboratory Study of Annotation Review

A second exploratory study assessed the use of CoScribe during review after class. In this setting, time is less scarce than during a lecture and learners can make use of the system's entire functionality. We evaluated the use of the CoScribe viewer for collaborative activities and the combined use of paper and digital documents.

Method

We recruited nine students (7 male, 2 female) among the participants of the first study. Each participated to a single-user session which lasted about one hour. Participation was voluntary and no compensation was given.

The participant was given an Anoto pen, a twenty page printout of slides of an introductory computer science lecture and several Digital Paper Bookmarks (see Section 7.1). He or she was seated at a table with enough free space for the paper documents. A computer screen on the table as well as a keyboard and a mouse provided access to the CoScribe viewer.

The sessions were structured as follows: At first, the participant was trained for five minutes on how to use the CoScribe viewer and Digital Paper Bookmarks. In the following, we requested the participant to perform given tasks with paper and the CoScribe viewer. This comprised creating annotations and bookmarks on paper as well as using the software viewer to modify own annotations. Next, we evaluated the appropriateness and compared the multi-user and the single-user visualizations of shared annotations. For this purpose, the participant had the task to browse through shared annotations of other users and to find specific annotations. These shared annotations were made by other participants in the previous study during their computer science lectures. Finally, in order to assess the ability of digital paper bookmarks to support cross-media navigation, we displayed a random slide in the software viewer and asked the student to find the corresponding page in the paper stack using the bookmarks he or she had created before. After the experimental session, feedback was gathered with a questionnaire containing 64 closed and open questions. These covered general questions about the usability of CoScribe, more specific questions on the visualization of shared annotations, on bookmarks, on the paper layout, on the general behaviour in lectures and personal questions. Finally, we conducted a semi-structured interview to gather further qualitative feedback. The entire session was videotaped.

Results and Discussion

Participants reported that in their established practice without CoScribe, handwritten annotations are typically not shared with other students due to the large effort that would be involved. We asked the participants for what purpose they would use shared notes. Of the variety of answers provided, five users mentioned that they would read the comments made by specific students known to take good notes. Two users stated that notes of different users complement each other, since there is not enough time during a lecture to note all information of importance. Two other users stated to correct own notes with the help of others. In the questionnaire, all participants judged CoScribe to be very helpful for collaboration ($M = 4.9$ on a 5-point Likert scale, $SD = .3, N = 9$).

We evaluated the novel multi-user visualization of handwritten annotations. For displaying personal annotations, the multi-user view is equivalent to the single-user view because the symbols for shared annotations can be easily hidden. Figure 5.12 shows results that compare the single-user view with the multi-user view concerning shared annotations of other users. The participants judged the multi-user view as more helpful when seeking an overview of them ($M = 4.3, SD = .9$ vs. $M = 2.1, SD = 9, N = 9$). This result is highly significant ($T = -6.64, df = 10, p < .001$). They judged this view to be roughly equally helpful for finding a specific shared comment (Fig. 5.12). Participants also judged the multi-user view to be roughly equally clear as separate views ($M = 4.3, SD = .7$ vs. $M = 4.6, SD = .5, N = 9$).

To compare the efficiency of both modes, we measured the time needed by the participants to find an arbitrary question and a specific comment on a given slide for each mode. Figure 5.13 shows that in both tasks, students found them much more quickly using the multi-user view ($M = 1.1s, SD = .4$ vs. $M = 2.5s, SD = 0.5$ and $M = 4.4s, SD = 1.6$ vs. $M = 8.0s, SD = 4.8, N = 6$ after having removed obvious outliers). These data are thus in keeping with the subjective ratings, but can only be considered a qualitative indication due to the small sample size. They support the

5.4 Evaluation and Discussion

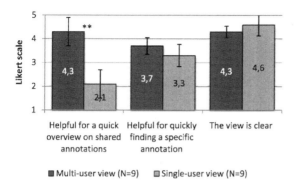

Fig. 5.12 Comparing the multi-user view and the single-user view (The error lines indicate the 95 % confidence intervals of the means. ** = statistically significant)

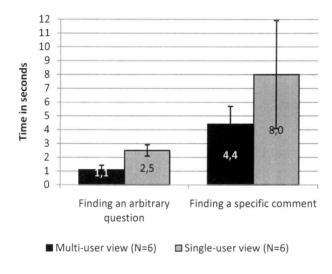

Fig. 5.13 Task completion times. (The error lines indicate the 95 % confidence intervals of the means.)

assumption that the multi-user view enables a fast overview of shared annotations of other users.

Participants found that the different icon sizes, with which annotations of other users are visualized and which reflect the size of the annotation, are helpful, but two participants had difficulties recognizing some of the smallest icons and suggested to visualize them in a more highlighted manner. All participants particularly valued the preview function, which visualizes all shared comments or all comments of a specific user in a down-scaled manner.

In the interviews, three participants stated that a list view of all annotations should complement the view to support users in systematically reviewing all annotations. One participant suggested a function for filtering out slides that are not important. This would allow for creating a printed summary of the lecture that contains only slides which are classified as important.

5.4.3 Study III: Performance of Handwriting Recognition

A substantial advantage of digital over traditional handwritings is the possibility to recognize handwritten text and to offer full text search within handwritings. In this section, we analyze the recognition accuracy for handwritten annotations and present an approach that significantly increases the performance for domain-specific terms.[2]

Method

For the evaluation, we used the Microsoft handwriting recognition engine that is part of Windows XP Tablet PC Edition [118]. It is an on-line handwriting recognition engine [119] that recognizes text from digital ink. In contrast to off-line handwriting recognition, on-line recognition uses not only the visual image of the handwriting, but relies on spatio-temporal data, i.e. the temporal sequence of the two-dimensional coordinates of the writing is available and used for the recognition. The engine utilizes a built-in dictionary of words of a given language, which can be extended or replaced by other dictionaries. It automatically segments a set of pen traces into individual words and separates text from graphics. For each word, a best guess and up to nine alternates with lower confidence scores are returned. The version available at the time of our experiments was not trainable to an individual user's handwriting. We used the German dictionary for our experiments.

First we evaluated the baseline performance of the handwriting recognition engine. We used a set of annotations in German language we had collected during the field study of CoScribe. From a set of 679 annotations that 10 students had made on lecture slides of a computer science lecture, we removed annotations that contained only drawings, whose text varied extremely in size (characters have more than 100 % of difference in height) or which were illegible to human readers. Of this reduced set, we randomly chose 169 handwritten annotations. The text of these annotations was manually transcribed.

Common metrics used to evaluate the performance of speech recognition and handwriting recognition engines are word error rate and character error rate. Both

---

[2] I gratefully acknowledge the graduate student Jie Zhou with whom I conducted this study in collaboration.

5.4 Evaluation and Discussion

metrics are based on the Levenshtein Distance [72]. This calculates the minimal distance between two strings by examining the minimal number of operations that is required to transform one string into the other. An operation can be an insertion, a deletion or a substitution of one single character.

The character error rate is calculated as follows:

$$CER = \frac{I+S+D}{N} * 100$$

where

- I is the number of inserted characters
- S is the number of substituted characters
- D is the number of deleted characters
- N is the maximum number of characters of both text strings

The word error rate (WER) uses the same formula, in which however the parameters I, S and D refer to operations on entire *words* and N is the maximum number of *words*.

Baseline Performance

As shown in Table 5.2, the resulting word and character error rates elevated at 49.2 % and 18.6 % respectively. These error rates are even higher than the (already unsatisfactory) rates reported in the literature: Koile et al. [65] used the same engine and reported a word error rate of 27 % on handwritten answers that students wrote on Tablet PCs as a response to specific questions of the instructor.

Our goal was to find out why in our case the error rate is that elevated. For this reason, we manually analyzed all 679 annotations that were made by 10 users in this lecture. The analysis showed that these annotations are much more complex to recognize than handwritten notes or answers. In contrast to answers that are submitted to the teacher, annotations serve a personal use. They have an informal character and are often written in a hurry. The annotations heavily varied in size, position and orientation. Even within one single annotation, the size of the characters can be very different (Fig. 5.14). Moreover, a considerable number of annotations (29 % of our test set) contained mixed text and drawings (Fig. 5.15). This is also due to to the topic of the lecture – sorting algorithms – in which students sketched many tree-like structures. Finally, annotations contain many (often personal) abbreviations as well as domain-specific terms (like $O(nlogn)$) or formulae.

Table 5.2 Baseline performance of the handwriting recognition of lecture annotations

	Percentage
Word error rate	49.2 %
Character error rate	18.6 %

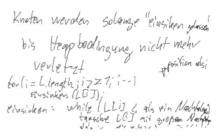

Fig. 5.14 Example of an annotation with text which is difficult to recognize

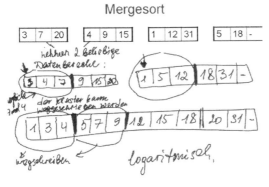

Fig. 5.15 An annotation that combines text with graphics

Improving the Accuracy of Handwriting Recognition

Given these problematic aspects, we aimed at improving the recognition accuracy of domain-specific terms. This is important because domain-specific terms are often used in search queries as well as for tagging and indexing documents.

For these experiments, we removed annotations that contain drawings from our test set. This results in a total of 118 purely textual annotations. A considerable percentage of 11.3 % of all terms in these annotations are domain-specific terms. The selection of domain-specific words[3] from the field of computer science and the

---

[3] The following terms, contained in the slides and/or the annotations, were classified as domain-specific (slides were in German): Pivot, Pivotelement, Worst Case, Sortiert, Quicksort, O(nlogn), O(n), O(n2), Zeiger, Daten, Baumstruktur, AVL-Baum, Schlüssel, Sortierung, Zwischenordnung, Durchläufe, Durchlauf, Speicherbedarf, Speicher, Binärbaum, Array, Suchbaum, Wurzel, Knoten, Sortieren, Sortierschritte, Mergesort, Heap, Heapbedingung, Heapsort, Laufzeit, Aufwand,Einsinken, tauschen, austauschen, Liste, Teilliste, optimieren, Sortieralgorithmus, Element, Java, Bottom-Up-Verfahren, Datenblöcke, Treesort, Out-of-Place, Sortierverfahren, Permutation, Datenmengen, Cluster, Inorderpartraversierung, Pseudocode, Datenstruktur, Rekursion, binär, Logarithmus, Heapbereich, rekursiv.

5.4 Evaluation and Discussion

classification of tokens was made by a domain expert (a fifth-year computer science student).

We examined three approaches. All of them used semantic knowledge to adapt the dictionary that is used by the handwriting recognition engine. They are based on the assumption that annotations made on lecture slides frequently contain words that are also written on these slides. In particular, this takes into account domain-specific words, which are typically not included in the standard dictionary.

Approach 1: Dictionary from all slides The first experiment adds the tokens from all slides of the given lecture to the dictionary. Hence, the dictionary is the same for all annotations from our corpus. It contains 2283 words taken from the 42 slides of the lecture. We will refer to this dictionary as Dictionary A.

Approach 2: Dictionary from current slides The second experiment uses different dictionaries for annotations made on different slides. The dictionary for a specific annotation contains all tokens extracted from the slide the annotation is located on. We will refer to these dictionaries as Dictionaries B.

Approach 3: Sliding-window dictionary The third experiment relies on a sliding-window approach. Again, annotations made on different slides have different dictionaries. The dictionary for a specific annotation contains all tokens extracted from the slide the annotation is located on and all tokens from the preceding five and the following five slides. If the slide is amongst the first or last five slides, the smaller number of all preceding or all subsequent slides is used. We experimented with different numbers of preceding and subsequent slides and found that in our case, the number of five slides provides the best results. As a matter of course, this number depends on the specific slide set. We refer to these dictionaries as Dictionaries C.

Improved Performance

Table 5.3 gives the recognition results for domain-specific words for all three types of dictionaries and contrasts them with the baseline. The use of either dictionary significantly reduces the word error rate. Dictionary B (all tokens from the current slide) clearly outperforms the other dictionaries. In contrast to using no domain-specific dictionary, the results show a relative word error rate reduction of almost 20 %. It is not surprising that the character error rate did not decrease, since the dictionary is not used for recognition on the level of individual characters.

Table 5.3 Performance of the handwriting recognition for domain-specific terms

	Word error rate (%)	Character error rate (%)
Baseline: No domain-specific dictionary	45.3 %	18.2 %
Dictionary A (all slides)	41.2 %	16.4 %
Dictionaries B (current slide)	**36.5 %**	16.2 %
Dictionaries C (sliding window)	41.8 %	17.4 %

Design Implications

The results of this evaluation show that handwritten annotations of lecture slides cannot be recognized with a high accuracy using a standard handwriting recognition engine.[4] Due to this low recognition accuracy, the interaction design of CoScribe does not require handwriting recognition. In particular, handwriting is not displayed as text but as the original handwriting. Nevertheless, handwriting recognition is used in the background. The text of each newly created or modified annotation is automatically recognized.

We have shown that the recognition performance for domain-specific terms can be significantly improved by using domain-specific dictionaries, which are automatically created from the annotated document. The recognized text can then be used for full-text search within the annotated document.

---

[4] Recently, Liwicki et al. [80] presented a similar approach for improving the recognition of handwritten annotations on paper documents, using the same handwriting recognition engine. Instead of text contained within the document, they extracted text from a personal knowledge base to improve the vocabulary of the recognizer. Their approach could reduce the overall word error rate by 4 %. They report a better baseline performance (approx. 30 % word error rate) than found in our results. We assume that this is due to the fact that their annotations contain only text; in contrast in our experiment many annotations contained sketches.

Chapter 6
Hyperlinking between Printed and Digital Documents

Knowledge workers frequently work with a set of interwoven documents, where information is distributed across several documents, e.g. in a printed lecture script, in some additional Web pages and in a personal notebook. In such interwoven documents, references are an important means that help the user in integrating this information. The following scenarios give some examples of how references support tasks in our application domain:

Scenario 12 (Integrating information from multiple documents) *In addition to some lectures, Sally attends a seminar. Her task is to compose a term paper on a specific topic. This requires mainly autonomous, self-directed work. The instructor indicates a few pointers to the literature. Based on these, she has to find further literature and precise the subject and the formulation of the questions. She searches literature in libraries, borrows books and photocopies some journal articles. She also looks for information on the Web and prints Web pages and PDF documents in order to have them available on paper. Initially this is a rather chaotic and unstructured collection of various documents. In order to structure the collection, Sally connects related documents by hyperlinks. A graphical overview on all documents and the hyperlinks between them is automatically created by CoScribe.*

Scenario 13 (Excerpting) *Excerpting is an important scientific method for reading books and summarizing their contents in a structured way. An excerpt is a short summary of the most important aspects of an existing text. This can also include own thoughts of the reader, can set the text into a larger context and can establish relations with previously acquired knowledge. While reading a text, Sally makes notes on a separate document, e.g. on a paper notebook. She adds references to the original text to indicate the passage a note refers to. This allows her to easily find the original passage at a subsequent point in time.*

Traditional handwritten references are difficult to follow, since the target document must be manually searched. This is particularly hard with cross-media hyperlinks. Moreover, users cannot easily share their references with co-workers. And even if references of co-workers are shared, it is not easy to understand their meaning because users have highly individual practices of referencing.

Table 6.1 Desiderata and our approaches concerning cross-media hyperlinks

Desideratum	Approach	Section	Concept	Function	Innovation w.r.t. related work
Easy and reliable creation of cross-media hyperlinks	Association gesture spanning physical and digital documents	6.2	•	•	•
Flexible scopes	Different types of association areas	6.2	•		•
	Physical folders		•		•
Integration of legacy media	Anoto-enabled registration stickers	6.2	•	•	•
Quick and easy following of hyperlinks	Tapping on link hot-spot	6.2	•	•	
	Same interaction in both worlds		•		•
	Cross-media history		•	•	•
Sharing hyperlinks	Button toolbar	6.3	•	•	•
Access to shared hyperlinks	Integrated multi-user visualizations	5.3	•	•	•
	Ecological view	6.4	•	•	•

In this chapter, we present a novel interaction concept for paper-digital hyperlinks. CoScribe supports users in adding own hyperlinks to existing documents in order to connect printed and digital documents in any combination. Similar to Vannevar Bush's early vision of Memex [12], the knowledge worker will be able to create a personal Web of information. In contrast to Bush, this closely connects physical with digital media. Cross-media hyperlinks are not only practical, being faster and easier to follow, but the conscious selection of passages to be linked supports successful learning. By translating contents into higher-level concepts and establishing relationships between them, learners build structural knowledge, which facilitates recall and comprehension and is essential to problem solving [51].

As discussed in Chapter 2, several approaches for cross-media hyperlinks have been presented in the literature. Our concept improves over the literature in several respects: Previous solutions require the user to switch between different interaction devices for paper and for digital media. Moreover, they do not integrate hyperlinks into the larger ecological context of a multi-person-multi-document environment. Furthermore, existing solutions do not address the collaborative use of hyperlinks. In contrast, our approach allows the user to create hyperlinks on both media with one single digital pen and one single pen-based association gesture. Like annotations, hyperlinks can be shared with other users. An ecological view provides integrated overview of and access to all hyperlinks that several users have made on a collection of documents. Table 6.1 provides an overview of our concepts.

6.1 Unified Pen-based Linking on Paper and on Displays

As discussed in Chapter 2, approaches proposed in related work require the user to switch between different input devices for interacting with physical and digital documents. Typically, the input device for interacting with printed paper is a digital pen. Interaction with digital media takes place using keyboards and mice or pen-based displays with styluses that are different from the digital pen which is used on paper. Hence, we encounter not only a discontinuity between physical and digital media, but also a *discontinuity of tools*: each tool applies either to printed or to digital media, but not to both of them. CoScribe supports using the same digital pen not only on printed media, but also on displays (Fig. 6.1, left and center). On paper, the Anoto pen acts as a traditional ballpoint pen and in addition captures the ink traces in an electronic form. On a display, the digital pen behaves similarly to styluses of common pen-enabled displays, such as those of Tablet PCs, with the difference that several pens can be simultaneously used on the same display. Almost all interactions can be made in a similar manner in both conditions. For instance, when creating a hyperlink between a printed document and a Web page, the user associates both documents with the *same* digital pen (Fig. 6.1, right).

In order to allow pen input on displays, we developed two Anoto-enabled display prototypes. A first prototype (Fig. 6.2 left) is particularly well-suited for individual work. It is of medium size with a display size of 82 cm and is slightly inclined for improving readability. A second prototype (Fig. 6.2 right) addresses co-located collaboration by multiple users. Its display size measures 112 cm. The technical setup of the prototypes is illustrated in Fig. 6.3. The displays use back-projection with a full HD resolution. During our experiments, we were using an Optoma HD 80 projector. To enable pen interaction, the image is projected onto a specific foil which the Anoto pattern is printed onto. Following the approach of Brandl et al. [10], we used

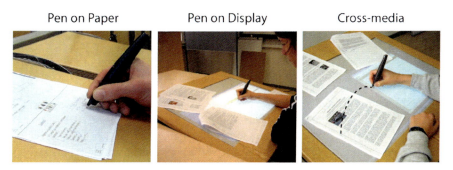

Fig. 6.1 Using the same digital pen to interact on paper and on displays

Fig. 6.2 Two prototypes of Anoto-enabled tabletop displays

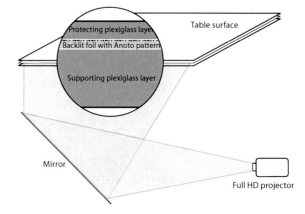

Fig. 6.3 Schematic illustration of the technical setup

a HP Colorlucent Backlit UV foil and printed the Anoto pattern onto it. This foil is put between a supporting plexiglass layer of 5 mm width and a layer of 1 mm width, which protects the surface. Recently, Hofer and Kunz [38] demonstrated a similar approach, based on a different foil, that enables to convert not only projection-based displays, but any standard LCD screen into an Anoto-enabled display.

Since we reserve a particular region of the Anoto pattern space for the display, the system can distinguish between pen interaction that occurs on paper and pen interaction that occurs on the display. The Anoto coordinates of pen events originating from the display are automatically converted to screen coordinates. If the event is located on a window of CoScribe, the system directly processes the event. Otherwise, a standard mouse event is generated. This enables controlling third-party applications with the pen.

A limitation of the current prototypes is that the Anoto pens leave slightly visible ink marks on the protecting layer, which however can be easily wiped off. A next version of the prototype should include an ink-repellent surface. Alternatively, future models of Anoto pens could control the ink cartridge and emit ink only if used on paper.

6.2 Creating and Following Hyperlinks

We model a hyperlink as a binary association between to document entities. It is symmetric and can be traversed in both directions. These two-way links follow the original hypertext principles introduced by Bush [12] and Nelson [102] and stand in contrast to the reduced model of one-way hyperlinks known from the World Wide Web.

In the early design phase, we investigated how students and colleagues make traditional handwritten references in documents. We observed that these are typically not made within the text but in the left or right margin of the document, leading to clearly visible references. Inspired by this practice, we opted for a spatial differentiation between the actual document and association areas. In these rectangular areas beside and above the document, users can create and follow hyperlinks. Association areas are not only printed on paper documents, but also displayed in digital documents.

A key issue of our interaction design is the separation between a *generic association gesture* and *different association areas*, which act as end points for the gesture and represent different scopes. This combines intuitive interaction with versatility and predictability even with the restricted feedback capabilities of a Pen-and-Paper User Interface.

Our approach relies on pen gestures for several reasons. First, the user makes annotations with a digital pen. Using the same pen for linking tightly integrates both interactions. Second, in contrast to gestures performed with fingers or hands or manipulations of the physical shape of a document, no further tracking infrastructure than the digital pen – which is already used for annotations – is necessary. While at a first glance, multi-touch seems compelling for associating two link anchors in one single step, our experiences showed that users typically first select a first link anchor and then seek for the appropriate second position of the association. This is better served by an interaction which is divided into two steps. Finally, pen gestures are very flexible. It is possible to mark and link small passages, but also entire documents. Pen gestures can be made both on paper and on pen-enabled displays. In contrast, linking documents by physically bringing them together is intuitive for entire documents but not for specific sub-passages within a document. Moreover, moving or collocating digital documents requires a different interaction than moving or collocating physical documents.

In the following, we first present the pen gesture for creating hyperlinks and then provide more details on the different types of association areas this gesture is performed on.

Association Gestures

Similar to the interaction technique for annotating documents, our technique for creating hyperlinks is designed to be minimally invasive and highly compatible with existing practices of creating references. The starting point are informal handwritten

132 6 Hyperlinking between Printed and Digital Documents

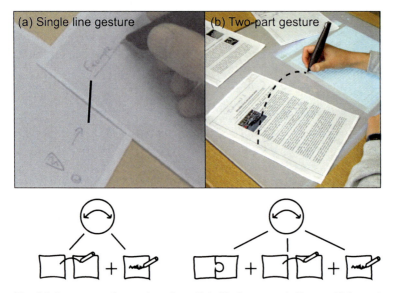

Fig. 6.4 Pen gestures for creating a hyperlink. The lower part indicates which core interactions are combined

references (e.g. "see p. 4", "cf. Wikipedia Digital Paper"), which indicate the link target in a way that can be interpreted by a human reader. This informal way of referencing is maintained. In order to additionally enable interpretation by computers, the user performs a quick and reliable pen gesture. Accounting for the associative nature of hyperlinks, this is an association gesture which connects two association areas.

Since one of our key design aspects is supporting cross-media linking, this gesture typically spans two different physical documents or associates a physical document and a digital document on a display. Our interaction model considers both a paper page and a screen as similar display instances. This interaction is therefore an extension to prior stitching gestures that span different paper documents [122] or several displays [37].

Using one single pen, the gesture is performed on paper, on the pen-sensitive display or on both of them. Two variants of the pen gesture exist. Figure 6.4 gives a visual overview while Figure 6.5 provides details on the gestures.

The first variant is a single-line gesture. In order to define the first link anchor, the pen is held down on a first association area for 500 ms without moving. If the system is used in streaming mode, a click sound is then played to provide feedback. Next, without lifting the pen, the user draws a line to the second association area, where he or she finally lifts the pen (Fig. 6.4 a and Fig. 6.5 lower left). This gesture combines the core interactions of *inking*, *combining* and *bridging* introduced in our interaction model in Chapter 3.

Alternatively, if the two areas are not close to each other, the user performs a two-point gesture. To do so, he or she makes two consecutive pen taps on both link

6.2 Creating and Following Links

Fig. 6.5 Details of the pen gestures for creating links

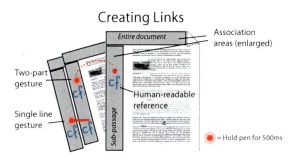

anchors instead of the connecting line (Fig. 6.4 b). This has the disadvantage that the interaction is not modeless, but on the other hand, the interaction does not require repositioning documents. This is for example necessary for links within different pages of the same book. Moreover, the user can start the link creation already before having selected the exact link target. This gesture combines the core interactions of *inking* and *bridging*.

The requirement of dwelling the pen for 500 ms permits a clear distinction between annotations and link commands. Moreover, it avoids confusing link start and link ending in the two-point gesture. If a computer is nearby, the gestures are recognized in real-time and instant audio feedback is given. A link can be deleted with a cross-out gesture on any marking made for creating this link.

This associative gesture can be quickly and easily performed during the work process.[1] Moreover, it does not rely on a possibly error-prone recognition of special keywords or symbols. This is particularly relevant as most current pens are not able to provide system feedback.

Association Areas for Defining Scopes

It is established practice to create references between passages of different extent. This includes referencing an individual figure or a short paragraph, entire pages or

---

[1] Related research [73] compared several pen-based techniques for switching between modes, one of them is tap-and-hold. The authors found tap-and-hold to be slower and more error-prone than switching between modes by varying the pressure or by using additional buttons. We nevertheless opted for tap-and-hold for the following reasons. First, pressure-based mode switching is not possible in paper-only environments, as it would be important to provide real-time feedback to the user about the current pressure level. Second, other techniques which require additional hardware were not acceptable in our case, as they heavily restrict deployability in real settings. Third, while tap-and-hold requires the user waiting a short moment before making the actual gesture, time is not critical in our case, since link creation is an infrequent activity. Finally, the main problem of tap-and-hold reported was that users had difficulties in holding the pen at a fix position on the slippery display of a pen-sensitive screen. This was no problem with CoScribe, as paper is not slippery and the screen is large enough to allows for some minimal movements. Our design decision was confirmed in the evaluation. Users had no difficulties in creating hyperlinks and reported to appreciate the clear distinction between a normal mode and the gesture for creating a hyperlink.

chapters, a whole book or even several documents. To account for this practice, our model supports the entities to be linked to be of different extent. An entity can have either of the following scopes:

- an entire document
- an area within a document (this area can span several columns and/or pages), or
- a collection of documents

Each document or document collection can be represented on paper or on the screen. Hence, associations can span the paper/digital boundary. While we model passages within documents as regions of space, our interaction technique could be coupled with the automatic extraction of document elements [165]. This would allow for linking specific semantic objects within the document.

The generic pen gesture described above has different meanings depending on which area it is performed on. This provides for an intuitive and flexible way of defining different link anchors. Different types of association areas are contained on printed and digital documents, on physical books and on physical folders. Each type of area represents another scope. Figure 6.6 provides an overview of these types of association areas. It shows how the physical areas are complemented by digital representations. In the document viewer, the areas are displayed at the same positions on the documents as on paper. In the ecological view (see Section 6.4 below), the nodes of the graph that represent the documents act as digital association areas. On these areas, the pen gesture for creating hyperlinks can be performed in the same manner as on paper. We now discuss each type of association area in turn.

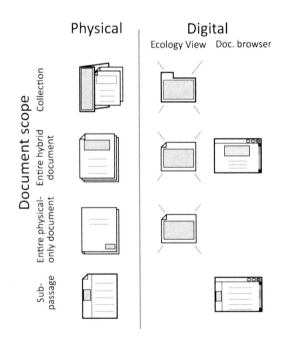

Fig. 6.6 Types of association areas

6.2 Creating and Following Links

Fig. 6.7 Links from and to physical books or journals are enabled by a small sticker which is covered with Anoto pattern

Entire Document Each document contains an area on the top of its first page where gestures are made for links that apply to this entire document (Fig. 6.5 top). Since one of our key design aspects is supporting cross-media linking, these association areas are included in each printed document page, in their digital representation in the document viewer and on Web pages in Mozilla Firefox. CoScribe therefore includes a Firefox extension which manages and displays areas and pen-based hyperlinks on Web pages.[2]

Entire Physical-only Document In order to support interactions with Anoto pens on physical-only documents (books, journals etc.), which do not contain the Anoto pattern, users can attach a small sticker covered with the pattern onto the document cover (Fig. 6.7) and register this sticker with the document's barcode. Depending on the pen technology, the user can directly scan the barcode with the digital pen or she manually enters the barcode. The digital metadata of this document (including an image of the cover, if available) is then automatically retrieved from a database. Our current implementation uses the Amazon.com web service.

Sub-document Level A margin area contained on each page of a document (apart from physical-only documents) provides for creating links from or to sub-passages within a document. It is included in printed documents, in the CoScribe viewer for digital documents, and in Mozilla Firefox. The user can define the precise extent of

---

[2] The Firefox plug-in is realized in JavaScript. The script is executed by Greasemonkey (http://www.greasespot.net) on each Web page which is accessed in Mozilla Firefox. The underlying principle is that our plug-in is lightweight and manages only the 'view' and 'control' elements of the Model-View-Controller paradigm. All information is managed by CoScribe. The plug-in therefore communicates with a CoScribe Firefox Service, which is implemented in Java and has access to the repository. The communication is performed over the MundoCore middleware [3] using a MundoCore plug-in for Firefox [126], which was developed at our institution. The plug-in displays the Web page in an IFrame, maintaining the normal visualization of and interaction with Web pages. In addition, the plug-in visualizes association areas for CoScribe hyperlinks and tags and visualizes existing links and tags. CoScribe tracks the pen interaction on these association areas. If a new link or tag is created or a link hot-spot is tapped on, the CoScribe Firefox service informs the plug-in about new contents or requests to display a new Web page.

Fig. 6.8 Physical folders provide for defining collections of documents in an intuitive manner

the linked document passage. She therefore draws a vertical line besides the passage where the association gesture is made. Such scope markings are deleted the same way as association gestures by a zig-zag gesture. If the document contains several columns, each column has an own association area. A scope can span several columns and/or pages.

In a user study on how people make references within traditional paper documents, we found that users often do not explicitly define the scope of a reference. Instead, the scope remains fuzzy and refers to an unspecified passage or item near the reference. In contrast to prior work, our interaction also supports fuzzy scope definitions. These associations are made in the association column without any further specification. In addition to these formalized gestures that are computer-interpreted, users can make any informal markings for the scope definition, which are displayed like normal handwriting in the electronic viewer (e.g. brackets, lines and arrows). While users understand them, the system is agnostic of their meaning.

Document Collections For links from or to a collection of documents (one-to-many or many-to-many), CoScribe includes the concept of folders. Physically placing a document into a folder is a very intuitive established interaction for defining a collection of paper documents. Users can then easily reference the entire collection by referring to the folder. Our slightly formalized interaction design is inspired by this practice. A folder (Fig. 6.8) contains one or several documents and can be the starting or ending point of an association gesture, as it contains an association area on its cover. Similarly, the digital representation of a folder in the ecological view acts as an association area. In order to detect that physical documents are added to a folder or removed from it, the interaction technique relies on location tracking of document and folder positions. At places where no tracking infrastructure is available, the user can alternatively perform a pen gesture to inform the system about an

6.2 Creating and Following Links

added or removed physical document. The same gesture applies if a user wants to add or remove a digital document.

Summing up, the conceptual activity of creating links combines up to three core interactions of our model. These are *inking*, *combining* and *bridging*. It moreover illustrates the principle of *spatial multiplexing*: the generic association gesture has different meanings depending on the area it is performed upon. Finally, *gestural multiplexing* enables to differentiate between creating associations and defining sub-document scopes.

Following Hyperlinks

A hyperlink is followed by tapping on or near a link marking (the gesture or the human-readable reference) in an association area (Fig. 6.9). This is possible on a printed document page, on a digital document on the display, on a book and on a folder cover.

The target document is then displayed in the CoScribe viewer or in Mozilla Firefox. The passage the hyperlink is referring to is highlighted in yellow. If the target is a folder or a physical-only document, it is displayed in the ecological view.

Backward and forward buttons support navigating in a cross-media history that contains not only Web pages as traditionally known from Web browsers, but also the digital representations of printed documents.

It is important to note that following digital hyperlinks always leads to a digital representation, regardless whether the source document is a printed or a digital one. This provides for visually skimming the contents of the target scope. If the user prefers to engage with the document more deeply (e.g. by active reading), he or she either can print a paper version or is indicated the location of an already existing printed representation (see Section 4.3).

Fig. 6.9 A hyperlink is activated by tapping on the link marking

6.3 Sharing of Hyperlinks

As with annotations, CoScribe supports sharing hyperlinks with other users. The default setting is that hyperlinks are visible to members of the user's group. If the user wants to set a different visibility, he or she can use classification buttons introduced for annotations in Section 5.2. Shared hyperlinks are made available in the document viewer, in the ecological view and in subsequent printouts of a document.

In the document viewer, shared links of other users are displayed in a similar manner as shared annotations (see Section 5.3 above). In addition to the annotations of a particular user, the single-user view includes the hyperlinks made by this user. In the multi-user view, the symbols of shared hyperlinks are displayed in a down-scaled manner. Similarly to shared annotations, shared hyperlinks can be copied into the own script or can be removed from the view. The visualization of shared hyperlinks in the printed representation follows the same principle as the visualization of shared annotations.

6.4 Ecological View

The document viewers display a visualization that corresponds to an *individual document*. However, as shown in Section 1.1, one of the main advantages of paper is that one can lay out multiple sheets and documents on the desk. This provides for a better overview of documents and their relations. We designed a digital visualization that corresponds to an augmented view of the *entire desk*. This so-called ecological view integrates all documents into one single interactive visualization. Its focus is on the relations between documents and on the activities that users performed on them.

This view is a digital equivalent to the arrangement of multiple documents in the physical workspace. However, in contrast to related work (e.g. [167]), it does not aim at mirroring the *physical* arrangement only by one. Instead, it details on the *logical* relations between documents. Documents are not visualized near each other because their physical locations are close, but because they are connected by a hyperlink, tagged with the same tag concept or used by the same person. This provides for a structured access to the information of document collections. The following scenario exemplifies how the ecological view can be used:

Scenario 14 (Structured Access to Information) *During the last few weeks, Sally has frequently met her learning partners. Together, they have revised the lecture handouts, have read many further documents and have written some new documents. Now that the final exam is arriving, Sally wants to review all information which is related to Molières comedies, one of the topics of the exam. She selects the tag "Molière" and has direct access to all information that was tagged with this label. Moreover, she seeks a document of which she does not remember the contents. But she knows that Dan has extensively worked on it during one of their first meet-*

6.4 Ecological View

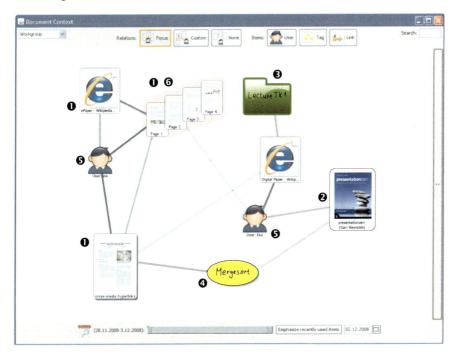

Fig. 6.10 The ecological view

ings. She therefore selects all documents which Dan has worked on and restricts the time period. Of the three remaining documents, she quickly finds the one she was looking for.

Due to the focus on relations, we opted for an interactive graph visualization. A screenshot of the view is given in Fig. 6.10 and will be explained in the following. Documents and document pages are represented by thumbnail images. Folders, tags and users are represented by icons. These nodes are connected by edges that visualize links, tags and user activities.

- *Anoto-enabled documents* (Fig. 6.10 (1)) are displayed with a thumbnail image of their first page and have a shadow that corresponds to their number of pages. For a more detailed view, users can expand a document to thumbnails of its individual pages (Fig. 6.10 (6)). Double clicking/two pen taps on a document or page thumbnail opens it in the corresponding document viewer.
- *Printed-only documents* (2) are visualized by a thumbnail of the cover.
- *Folders* (3) are indicated with an icon in their color containing a handwritten label. A shadow corresponds to the number of pages of all documents contained in the folder. Folders can be "unfolded" to view the individual documents.
- *Tags* (4) are displayed as oval concept cards covered with the tag label.
- Each *user* (5) is represented by an icon.

An edge expresses:

- A *link* between two documents or folders. When clicking on the edge, the scope of the link is highlighted in yellow in the document thumbnail(s).
- *Tags* of documents or folders are indicated by an edge between the tag concept card and the document or folder.
- *User activities* are expressed by an edge from a user icon to all documents this user has annotated, linked or tagged.

For example, the graph provides a quick overview of how the collection of documents is structured by hyperlinks, of which user has worked on which documents or of which documents are tagged with a specific tag concept.

A collaborative graph which contains the documents and activities of several users can quickly become complex. This makes high demands both on navigation and on appropriate reduction of the complexity. For this reason, the interactive visualization enables users to navigate through a large graph by panning and zooming as well as by collapsing and expanding documents and folders. Moreover, the ecological view offers comprehensive filter options on five dimensions:

A dual-end time slider (Fig. 6.11 (1)) acts as a temporal filter. It enables to hide all items that have not been created or modified within a given period. To further allow a quick temporal overview of all items, a time mode visualization varies the size of the nodes. Larger nodes represent items that have been created or modified recently while smaller nodes correspond to older items.

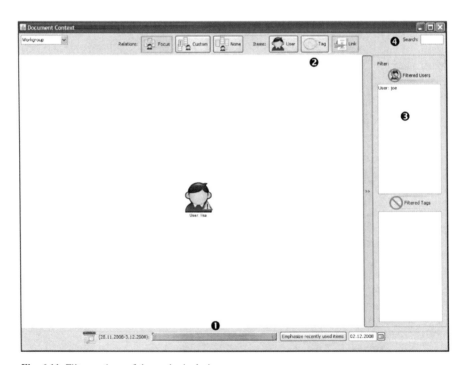

Fig. 6.11 Filter options of the ecological view

Toggle buttons allow the user to hide or show all links, all tags or all users (Fig. 6.11 (2)).

A further filter allows for hiding an individual user or an individual tag concept: All documents which are modified only by this user (or tagged with only this concept) are hidden. This filter is activated by dragging the corresponding node and dropping it on a list of hidden elements which is displayed in a filter panel to the right of the graph (Fig. 6.11 (3)). The user or the tag is then added to this list. By tapping or clicking on it, the documents are displayed again. It is important to note that not only one user or one tag can be filtered, but an arbitrary number of them.

It is also possible to focus on one node. Only this node and all directly adjacent nodes are displayed. All other nodes are hidden from the graph. They are listed in the filter panel and can be re-displayed by tapping or clicking on one or more entries in the filter panel.

Finally, the view supports full-text search. An arbitrary query string can be entered in a text input field (Fig. 6.11 (4)). All users, tags and documents that do not contain this string are hidden from the view.

6.5 Evaluation and Discussion

In a user study we evaluated how CoScribe supports users in integrating information which is distributed between several printed and digital documents. This setting was inspired from the findings of one of our field studies, which show that knowledge workers frequently use interconnected collections of printed and digital documents.

We assessed the ease-of-use and the learnability of the interactions for creating and following cross-media hyperlinks as well as the use of the pen-enabled tabletop display. In addition, we had the following hypotheses for the experiment:

H1: Cross-media hyperlinks enable the user to perform a complex information retrieval task in a set of interlinked printed and Web documents more quickly than with traditional pen and paper.

H2: Printed and digital documents are perceived as being more closely connected when using cross-media hyperlinks.

Method

A total of 10 psychology and 6 computer science students (9 female, 7 male) participated to single-user sessions of 90 minutes length. Their average age was 25 years (SD=5.5). The participants judged their computer skills as high (M=5.5 on a 7-point Likert scale, SD=1.5). They are experienced knowledge workers indicating to work 24.0 hours per week with digital documents (SD=10.9) and 13.5 hours with paper documents (SD=7.6). A considerable amount of time (M=11.2 hours, SD=7.9) is spent on working simultaneously with both types of documents. Participation to the study was voluntary and no compensation was given.

The participants used an Anoto pen, paper documents and digital documents on the pen-enabled display. Eight participants used the display in a tabletop configuration while the remaining eight participants used it as a vertical screen.

The sessions were structured as follows: After three minutes of training, the first task was to create arbitrary hyperlinks between a printed document, which was provided by us, and Web pages. The next task was to follow these hyperlinks. These initial tasks served for getting experienced with the system.

The remaining time of the experiment consisted of answering questions on historic murder cases. For this purpose, the participant was given a collection of printed and Web documents that contained information about this case. The information was collected from newspaper articles and diverse Web pages. This collection was pre-structured (annotated and interlinked) with respect to specific questions. To find the answers to our questions, the participants had to quickly get an overview of the documents, to find relevant passages and to read these passages in detail. During the experiment, we observed the participants navigating within the document collections and measured the time needed until the questions were correctly answered.

A within-subject design was used for this experiment. The participant was trained on a first document set. Two other document sets were then used for testing under either condition. We counterbalanced the document sets and the order of the two conditions. In one condition, the participant could use CoScribe. Links on printed documents could be followed by tapping with the pen on the hyperlink. Digital documents could be accessed on the pen-enabled display. In the control condition, the participants used traditional pen and paper and a traditional Web browser. The printed documents contained the same annotations and hyperlinks as in the other condition, but hyperlinks were expressed as handwritten references pointing to the number and the page of the target document. The participants were given an overview document that listed all documents, their numbers and the URLs of Web pages. References on Web pages were visualized by the Web annotation tool Diigo[3]. They indicated the number and the page of the target document. Moreover, target passages of references were highlighted in yellow, as done automatically by the CoScribe Web browser plug-in in the other condition.

Each document collection consisted of 3 to 4 printed documents with a total of 15 to 16 A4 pages that mainly consisted of text. Moreover, it included 3 to 5 web documents from Wikipedia and online newspaper archives. The documents comprised a short biography of the murderer, detailed documents on the murder and some more documents with background information on the case. The questions asked for factual information. Nevertheless, a correct answer required to integrate information from various passages of different documents. The relevant information for answering a question was distributed between 3 to 8 passages. Each collection contained between 8 and 10 hyperlinks that connected related passages. In order to guarantee comparability across the participants and for reasons of experiment duration, we opted not to let the participants themselves prepare the document collections.

---

[3] http://www.diigo.com

Instead, we asked a student who was not familiar with our system to prepare the document collections for all participants.

With this task, we aimed at finding out if the participants would be able to complete a realistic information integration task more quickly using CoScribe. Obviously this task goes far beyond simply following a hyperlink (which obviously takes less time than manually searching the referenced passage). Users had to handle many documents containing not only relevant, but much irrelevant information. Moreover, the users had to decide on the relevance of hyperlinks, since only about one third of all hyperlinks linked to passages which were relevant for a specific question. Finally, the participants were pressed for time (15 min. per document set for three questions). This setting thus represented a realistic knowledge work task where a co-worker had pre-structured a collection of documents with regard to specific aspects.

After the experimental session, we gathered subjective feedback with a questionnaire. This contained 24 items covering the usability of cross-media hyperlinks and of the pen-enabled display, the perceived proximity between paper and digital documents and personal information. For the sake of readability, we will present all statements in their positive form, but the form varied in the questionnaire. Finally, we conducted a semi-structured interview with each participant to gather qualitative feedback.

Results and Discussion

Creating and Following Links After a few minutes of training, all users had learnt how to use the pen on the display and how to create and follow links. Responses to an open question showed that the interaction was appreciated as being "easy" (8 persons), "quick" (2 persons) and "highly intuitive" (2 persons). We observed that all but one participant navigated through the interlinked documents with ease and confidence as if they had been using the system already for a long time.

In the responses to the questionnaire, creating hyperlinks was judged to be significantly easier and faster with the system than in the control setting (Fig. 6.12). Creating a hyperlink was rated to be very easy[4] in contrast to the value for creating traditional references which was below average[5]. Alike, creating a cross-media hyperlink was rated to be significantly faster[6] than creating a traditional reference[7].

In the interviews, the users valued the tap-and-hold gesture for starting hyperlinks, as this permits a clear differentiation between annotations and the link command. The gesture was considered to be very easy and quick to perform. This stands in contrasts to the findings of Li et al. [73] who report higher error rates and larger task completion times with hold-down gestures. This contradiction might relate to

---

[4] $M = 6.6$ on a 7-point Likert scale, $SD = .8, N = 14$
[5] $M = 3.0, SD = 1.5, N = 14; T = 7.66, df = 13, p < .001$
[6] $M = 6.5$ on a 7-point Likert scale, $SD = .7, N = 14$
[7] $M = 3.3, SD = 1.7, N = 14; T = 7.19, df = 13, p < 0.01$

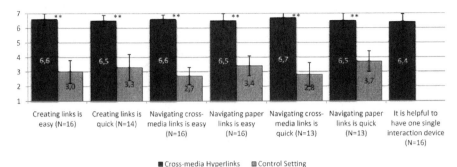

Fig. 6.12 Quantitative responses for hyperlinks (all statements are transformed to their positive form. Error lines indicate the 95 % confidence intervals of the means. ** = statistically significant)

the fact that Li et al. used tap-and-hold for mode switching in a drawing application, where mode switching occurs very frequently. In contrast, creating links is an activity that is performed only at an infrequent basis. In this latter case, holding down the pen for half a second is much more acceptable. As indicated by several participants, this short interruption is even supportive because it underscores that a special mode for creating (instead of following) a hyperlink is now being entered.

The separate hyperlink column was appreciated for enabling a quick overview of all links. However, four participants desired the possibility to also create hyperlinks on individual words directly within the document.

As far as following links is concerned, this was also rated to be significantly quicker and faster than in the control condition (see Fig. 6.12). The difference is particularly pronounced for cross-media hyperlinks between a printed document and a Web page, but also for hyperlinks between two paper pages, the difference is considerable. Following cross-media hyperlinks was judged to be very easy with CoScribe[8] whereas the value is significantly lower in the control setting[9]. Following hyperlinks between two paper documents was judged to be slightly less easy[10], but still much easier than without CoScribe[11]. Again, this difference is statistically highly significant.[12] A cross-media hyperlink can be followed significantly more quickly[13] than a traditional reference[14]. The difference is slightly less pronounced, but still significant, for links between two paper documents with CoScribe[15] vs. in the control setting.[16]

---

[8] $M = 6.6$ on a 7-point Likert scale, $SD = .6, N = 16$
[9] $M = 2.7, SD = 1.2, N = 16; T = 10.7, df = 15, p < .001$
[10] $M = 6.2, SD = 1.2, N = 14$
[11] $M = 3.4, SD = 1.9, N = 14$
[12] $T = 4.26, df = 13, p = .001$
[13] $M = 6.7, SD = .8, N = 13$
[14] $M = 2.8, SD = 1.5, N = 13; T = 5.31, df = 12, p < .001$
[15] $M = 6.5, SD = .9, N = 13$
[16] $M = 3.7, SD = 1.3, N = 13; T = 5.31, df = 12, p < .001$

6.5 Evaluation and Discussion 145

It was considered very helpful to have the same interaction device for printed and digital documents.[17] Participants reported that this made the interaction faster and connected printed and digital documents more directly. In the interviews, 14 participants valued that they did not have to switch between different devices for paper and for the computer screen. One participant emphasized on the fact that using the same pen for creating a hyperlink between paper and digital documents is helpful, as it gives the feeling of connecting documents more directly.

Pen-enabled Display When comparing the pen-enabled tabletop display with the pen-enabled vertical screen, the tabletop configuration was clearly preferred. While four participants reported in the interviews to prefer a vertical screen because it is easier to read and printed documents cannot cover parts of the display, ten participants preferred a horizontal tabletop configuration. This is inline with the findings which Morris et al. reported on active reading with different display configurations [99]. Most important reasons mentioned in the interviews were first that it is more natural and ergonomic to use a pen on a horizontal surface and second that printed and digital documents are more closely coupled using one surface for both of them. However, three participants perceived an extra effort for rearranging documents on the display. This discomfort is due to two deficiencies of our current prototype: the space provided on the tabletop is limited and no provision is made for coping with occlusions of displayed pages.

A further problem of the current tabletop prototype is the image quality of the display. While this is comparable or even superior to the quality of other tabletop displays, is does not yet reach the quality of standard screens. While this is not problematic in many application areas of tabletop displays, text has high requirements concerning display quality. Both contrast and luminance of our current prototype are acceptable for using the display for some hours but must be improved if the display shall be used on a regular basis.

When asked about the convenience of using the pen on the screen, the participants who had used the screen in a tabletop configuration reported a better score[18] than those of the vertical display condition[19]. While this difference is not significant, it is in line with our observations that it was more difficult for the participants to use the pen on the vertical display.

Performance Gain As depicted in Fig. 6.13, completing the information answering task with CoScribe took in average only about 60 % of the time that was needed in the control setting. This difference is highly significant[20] and confirms our hypothesis H1. As the two document collections did not result in significantly different completion times, we analyzed them together.

Figure 6.14 gives the completion times for the individual questions. This shows that all questions have been solved more quickly using CoScribe except for the

---

[17] $M = 6.4$ on a 7-point Likert scale, $SD = 1.1, N = 16$
[18] $M = 6.3, SD = .7, N = 8$
[19] $M = 5.6, SD = 1.4, N = 8$
[20] $T = -3.22, df = 15, p < .01$

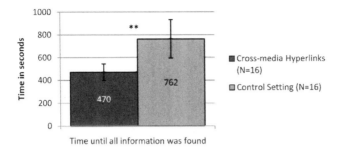

Fig. 6.13 Task completion times (error lines indicate the 95 % confidence intervals of the means. ** = statistically significant)

third question in the second document collection. Due to an inattention during the preparation of the document collections, one link of the CoScribe condition has been placed at an incorrect position. This made it more difficult for the participants to find the correct answer. The large time difference for the second question in the second collection attributes to the fact that in CoScribe all links are symmetric and automatically visible on both endings. In contrast, traditional handwritten references are not automatically visible at the target passage. This made it more difficult to find a specific information in the control condition.

We observed that it is very important to label a link anchor with some information about the target document (such as "Biography of the murderer"). While most links were labeled, each document collection contained the same number of unlabeled links. With the latter ones, the participants were much more likely to get disoriented and to be uncertain if they had already followed these links.

Gap Between Printed and Digital Documents In the questionnaire, we asked the participants as how closely connected they perceive printed and digital documents. Figure 6.15 shows that with cross-media hyperlinks and the pen-enabled display,

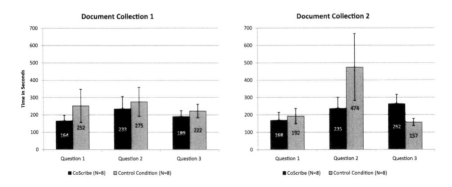

Fig. 6.14 Task completion of individual questions

6.5 Evaluation and Discussion

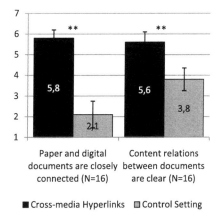

Fig. 6.15 Closeness of printed and digital documents (error lines indicate the 95 % confidence intervals of the means. ** = statistically significant)

printed and digital documents are considered to be significantly more closely connected[21] compared to the control setting[22]. Alike, the relations between the contents of printed and digital documents are found to be closer with CoScribe[23] than in the control setting[24]. These findings are in-line with the positive perception of the unified pen device and with the increase of performance found in the study. They suggest that cross-media hyperlinks effectively reduce the gap between printed and digital documents.

---

[21] $M = 5.8$ on a 7-point Likert scale, $SD = .8, N = 16$
[22] $M = 2.1, SD = 1.3, N = 16; T = 9.8, df = 15, p < .001$
[23] $M = 5.6, SD = 1.0, N = 16$
[24] $M = 3.8, SD = 1.1, N = 16; T = 5.2, df = 15, p < .001$

Chapter 7
Paper-based Tagging of Documents

Besides annotating documents and creating references, *categorizing* documents and *structuring the information space* are amongst the most relevant activities that knowledge workers perform when they seek to make sense of information. Structuring documents and collections of documents, translating them into higher-level concepts and establishing relationships between concepts is crucial for successful knowledge work. Tagging can transform an unsorted and possibly confusing collection of a large number of disparate documents to a unified and well-structured document space. Moreover, a well-structured document space fosters understanding and retrieval of information:

Scenario 15 (Creating concepts) *Sally uses tags to prioritize documents according to their relevance for her research questions. In a subsequent step, she systematizes the information: She first identifies abstract concepts. Then she tags each important passage with the appropriate concept. By doing so, she has not only understood the individual aspects but has acquired structural knowledge of the problem domain.*

This scenario gives only two examples of how tags can be used for successful learning. Tags support a variety of other activities. These include for instance structuring a domain with abstract concepts and relating these to document passages, marking up the structure of a document (e.g. by tagging the section headings), and prioritizing documents. Further activities comprise recommending documents to a collaborator (e.g. by tagging documents with his or her name), labeling a document with the physical place it is filed to easily find it at a later point in time, and structuring the temporal process of meetings.

This chapter presents pen-based and tangible interaction techniques for cross-media tagging. We define a tag as follows: A tag is a concept which is assigned to an entity, i.e. to an artifact or to a process. The concept is represented either by a predefined category *(category tagging)* or by one or several keywords which are freely chosen by the user *(free tagging)*. The same concept can be assigned to several entities. If appropriate, we will distinguish between the *tag concept* (a

Table 7.1 Desiderata and our approaches concerning cross-media hyperlinks

Desideratum	Approach	Section	Concept	Function	Innovation w.r.t. related work
Free tagging	Digital Paper Bookmarks, Tag Menu Card	7.1, 7.2	•	•	•
Category tagging	Digital Paper Bookmarks, Button Tagging, Process Cube, Process Knob	7.1, 7.3, 7.4	•	•	•
Full visual feedback on paper; easy physical access	Tagging with paper stickers	7.1	•		•
	Digital Paper Bookmarks			•	•
Collaborative structuring of documents & meta-cognitive support	Set of semantic types of bookmarks; collaborative view	7.1	•	•	•
Tagging with flexible scopes; convergence of tags; operations on the tag set	Defining tags on a separate paper tool	7.2	•		•
	Tag Menu Card			•	•
Quick tagging while annotating	Button Tagging	7.3	•	•	•
Tagging of temporal processes	Manipulation of tangible objects	7.4	•		•
	Exemplary model of phases		•		•
	Process Cube			•	•
	Process Knob			•	•

specific category or a given keyword) and the *instances* of this tag (the assignments of this concept to entities).

As discussed in Chapter 2, previous work on paper-based tagging relies on pen gestures and printed buttons. In contrast, we present a variety of tagging approaches that leverage to a higher degree the tangible character of paper. In the remainder of this chapter, we will present four interaction techniques for tagging. The techniques make extensive use of paper as a tangible tool that is inexpensive and can be easily produced by the users themselves. Digital Paper Bookmarks are based on paper stickers. The Tag Menu Card is a paper card that provides an inventory of tag categories. Button Tagging relies on interactive areas that are printed on paper and selected with the pen. Finally, we present two concepts of how tangible tools can be used to tag the temporal dynamics of collaborative processes. These four techniques

7.1 Tangible Tagging with Stickers: Digital Paper Bookmarks   151

complement each other, serving different purposes and having different properties. The choice of an appropriate technique depends both on the task and on the user's personal preferences. We conclude by an integrated discussion of all techniques. Table 7.1 provides an overview on the challenges which we address in this chapter.

7.1 Tangible Tagging with Stickers: Digital Paper Bookmarks

Paper bookmarks have proven to be an effective means not only for quickly accessing specific pages or parts of a paper document, but also for individually structuring a document. Impressive examples are books of law students, which often contain several dozens of bookmarks attached to the margins of the pages (Fig. 7.1 left). As a metaphor, bookmarks also quickly became familiar in the electronic world for marking documents on the Web. Empirical work shows that their use positively influences the perceived ease of finding information [152]. We aim at leveraging the ease of this interaction also for structuring *digital* documents via their printouts.

With CoScribe, we introduce Digital Paper Bookmarks, which span both paper and the electronic world. Digital Paper Bookmarks are adhesive stickers of different colors which can be attached to physical pages of printed documents at arbitrary positions (Fig. 7.1 middle and right). They are covered with the Anoto pattern and can therefore be labeled with a title using a digital pen. They are synchronized with the digital system and serve as digital bookmarks for these pages. Hence, Digital Paper Bookmarks combine the advantages of intuitive paper-based bookmarking with digital support. For example, bookmarks can be used for tagging the structure of a document by bookmarking the beginning of each section or for marking up important passages of a document.

  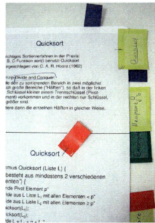

Fig. 7.1 Paper stickers are a powerful means for indexing documents (left). Digital Paper Bookmarks (center and right) take on this principle and serve as physical and digital bookmarks

The interaction for creating a Digital Paper Bookmark is inspired by traditional bookmarking practice. It involes three steps (see Fig.7.2):

1. *Combining:* A bookmark is attached to an arbitrary page of a printed document at arbitrary positions. Several bookmarks can be attached to the same page. The document has to be covered with the Anoto pattern.
2. *Associate (bridging):* With a simple pen gesture, the paper bookmark is associated with the page it is attached to. It is then also available as digital bookmark. This association gesture is a short line connecting the bookmark with the page it is attached to. An arrow is printed on the bookmark to indicate the pen gesture (see Fig. 7.3).
3. *Label (inking):* Since the Anoto pattern is printed on the stickers, one can use the digital pen to write a tag or a title on them, which will be synchronized.

The interaction thus combines three core interactions of our interaction model (see Section 3.4.2). The association step is not necessary if technology for tracking the location of paper sheets is used, which automatically detects the combination. The manual association lowers the technical requirements and enables bookmarking during mobile use. This proposed solution for creating bookmarks is entirely mobile and highly compatible with existing practices.

After creation, the position of a bookmark can be modified by attaching the bookmark to another physical position and performing the association gesture at this new position. Handwritten labels can also be modified. A bookmark can be deleted by physically removing it from the paper sheet and performing a cross-out gesture on the bookmark for deleting its digital counterpart.

A Digital Paper Bookmark has a dual function: It is an *instrument* for creating, modifying and deleting tags. Once created, the instrument also becomes a *first-class*

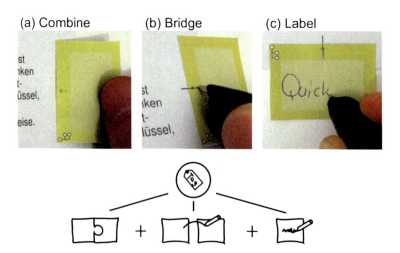

Fig. 7.2 Three intuitive steps allow the user to create a Digital Paper Bookmark. These correspond to three core interactions of our interaction model

7.1 Tangible Tagging with Stickers: Digital Paper Bookmarks 153

object, as the physical bookmark represents the actual tag. Since the bookmark is visible on paper, full visual feedback is available without additional digital support. Moreover, its shape provides a strong affordance for quickly accessing the physical page which has been bookmarked. This is a clear advantage over other classification means discussed in the literature (e.g. [75, 170]), which do not modify the physical shape of the document.

However, it is precisely this advantage that makes it difficult to re-print a document including the bookmarks. In order to address this problem with current technology, the print module prints a small image of the bookmark at each position of the document where a bookmark was attached in the original document. This is a visual indicator for the user who can then add bookmarking stickers at these positions. Clearly, this is a workaround which is not efficient when a document contains a large number of bookmarks. Future printers could solve this problem by automatically attaching bookmarking stickers at the appropriate positions while a document is printed.

Semantic Types

In addition to free tagging with keywords, Digital Paper Bookmarks offer predefined categories that support users in structuring documents. Users can choose between different types, each represented by a specific color and a specific symbol.

Offering different types of bookmarks has two main advantages: First, some common semantic classes facilitate the computer interpretation of bookmarks, the sharing with other users and automatic aggregations. Second, if bookmarks are used in a learning setting, the availability of specific types can support meta-cognitive learning processes by encouraging and reminding users to perform specific important learning activities which are related to bookmarking.

We developed a first set of semantic types to be used with Digital Paper Bookmarks in learning processes. It comprises four types. We distinguish the following four types on two layers:

The two semantic types on the *structure layer* support learners in structuring the learning documents: Section bookmarks (yellow) mark the beginning of a new

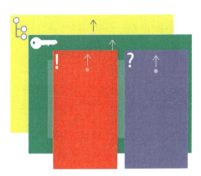

Fig. 7.3 The current version of CoScribe includes four semantic types of bookmarks

section. The title of this section can be written on the bookmark. Key bookmarks (green) can be used to mark key passages, where subject matters are defined or explained.

The two semantic types on the *meta-cognitive layer* serve for planning further learning activities as well as for controlling past activities. They can be used for assessing both the relevance of subject matters and personal difficulties. Important bookmarks (red) mark passages which seem particularly relevant to the learner. Unclear bookmarks (blue) point out passages which the learner has difficulties understanding and which therefore require further work. Once the subject matter is well understood, the bookmark can be removed.

The bookmarks of the structure layer offer much space for writing a title, whereas the meta-cognitive bookmarks serve as simple category markers without a title (Fig. 7.3).

We assume that the bookmarking actions keep easy and intuitive albeit the certain amount of abstraction imposed on the users by these four categories: On the one hand, marking documents with bookmarks of different colors is widespread in the paper world; on the other hand, the relatively small number of four categories still seems manageable. In order to help novel users memorize the meanings of the different colors, small symbols are printed on the paper bookmarks.

Visualizations

As Digital Paper Bookmarks are automatically synchronized with the electronic system, paper bookmarks can be integrated to and displayed within the corresponding digital document. CoScribe includes three different visualizations of bookmarks.

In the CoScribe viewer, bookmarks are visualized along with the document pages similar to how they appear on paper (Fig. 7.4). Clicking or tapping on a bookmark displays the bookmarked page. Hence, using bookmarks automatically creates a personalized index of contents of the document. This includes passage headings (yellow) as well as pointers to key passages (green), particularly relevant passages (red) and unclear passages (blue). In addition, bookmarks are also visualized in the slide preview panel.

A second visualization provides visual guidance for cross-media navigation. Depending on the specific task, users frequently switch between printed and digital representations of documents. For example, the user is likely to switch from the printed to the digital representation for viewing shared annotations that other users have made on this document. The user might switch in the reverse direction when she desires to deeply engage with the document and therefore prefers the printed representation. These switches require quickly finding the corresponding position in the other representation. Support for finding the digital document instance which corresponds to a physical instance has been developed in other research (e.g. [82, 135]) and is also included in our system.

The reverse direction is more challenging. Even if the location of printed documents is electronically tracked and can be indicated by the system, this provides

7.1 Tangible Tagging with Stickers: Digital Paper Bookmarks

no support for finding a specific page within a document. Digital Paper Bookmarks support this specific task by offering synchronized visual anchor points within both printed and digital document representations. A specific bookmark view in the software viewer displays a three-dimensional representation of the paper stack. It visually indicates the number of pages of the document, the current position within it and the bookmarks associated to it (Fig. 7.4 lower left). At a single glance, the user gets information about the approximate position of the page within the document, about the nearest bookmark and the page distance to this bookmark. The user can then find the paper page by recurring to its approximate position in the document or by choosing an appropriate bookmark located near this page. Furthermore, the bookmark panel also supports navigation within the digital document. Pages can be displayed by clicking or tapping on the three-dimensional representation of the paper sheet or on a bookmark. We decided not to integrate shared bookmarks of other users in these visualizations, as they would loose their key property of mirroring the bookmarks of the printed representation. This could be very confusing when using bookmarks for cross-media navigation.

A third visualization supports collaborative structuring. Through the contrasting of their own structure with the structure of others, with their markings of relevant or unclear passages, learners can assess and improve their own understanding of the material. Cognitive conflicts can arise and can lead to a modification of one's own bookmarks. This is supported by a third, collaborative visualization which focuses

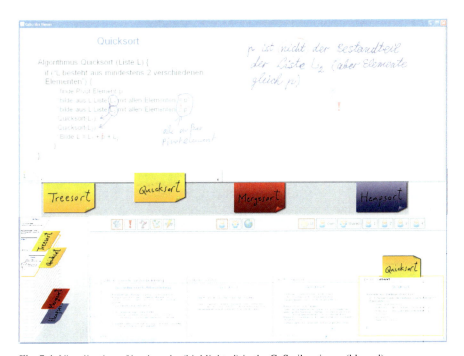

Fig. 7.4 Visualization of bookmarks (highlighted) in the CoScribe viewer (blurred)

on the structure of an individual document. It is depicted in Figure 7.5. The pages of a document are aligned vertically and represented in an abstracted manner to the left (Fig. 7.5 a). The user's own bookmarks are visualized beside (Fig. 7.5 b). The bookmarks of one or several members of the user's own group can be displayed simultaneously and are visible to the right of one's own bookmarks (Fig. 7.5 c).

In addition to this sharing within a small group of users, the bookmarking data of all users is automatically aggregated by the system. Anonymity is preserved as only the positions and types of the bookmarks are taken into consideration, but not their handwritten labels. An aggregated view visualizes data from all users indicating the type, position and frequency of bookmarks by colored markings of different sizes (Fig. 7.5 d). For instance, large red markers indicate that these passages are judged particularly relevant by a large number of learners and green rectangles mark key passages.

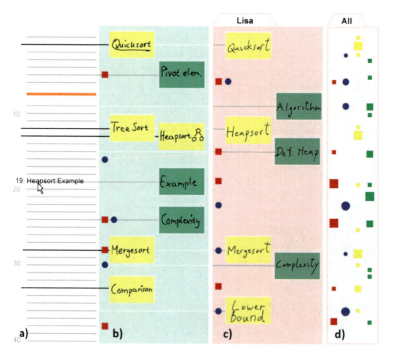

Fig. 7.5 A collaborative view for comparing the structuring of documents. It includes an abstracted representation of the document (a), own bookmarks (b), bookmarks of other members of the user's group (c) and an aggregated view of the bookmarks of all users (d)

7.2 Tagging by Association: Tag Menu Card

While Digital Paper Bookmarks is a very intuitive interaction technique, it is rather course-grained since bookmarks apply to entire pages. Moreover, only a restricted number of bookmarks can be attached to a document without the bookmarks becoming confusing. Therefore, we introduce a second interaction strategy for tagging documents. This relies on one or more separate paper cards for defining and applying keyword-based tags (Fig. 7.6).

Each Tag Menu Card contains several empty areas. At any time the user can define a new tag by writing one or several freely-chosen keywords in one of these areas. After a tag is defined, it is applied using any of the following interactions:

- Writing the tag on an association area of a document (as defined in Section 6.2) and enclosing it with a circle in order to mark it as a tag (Fig. 7.7 a). The tag is automatically recognized from the set of previously defined tags on the Tag Menu Card using handwriting recognition.
- Writing the tag on an association area and additionally performing the pen gesture for hyperlinks (Section 6.2) to associate it with the corresponding area on the Tag Menu Card (Fig. 7.7 b). This small additional effort ensures that tagging is correctly performed, as it does not rely on handwriting recognition. This is important when no computer is nearby, as most current digital pens cannot provide instant feedback on success or failure of the recognition.

Although the Tag Menu Card has a printed representation, it can be used for tagging both printed and digital documents. The precise scope of the tag within the document is defined the same way as when creating hyperlinks (see Section 6.2). Depending on the type of association area where the tag is written on or associated

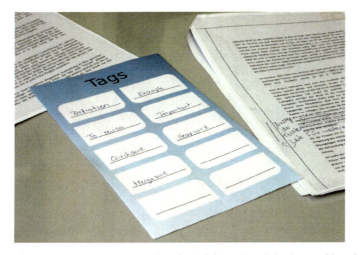

Fig. 7.6 A Tag Menu Card provides for defining and applying keyword-based tags

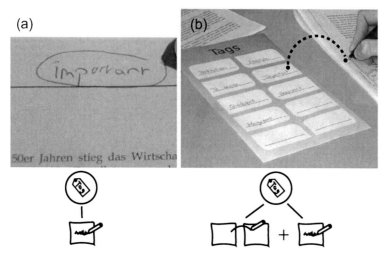

Fig. 7.7 Two interactions for applying tags (that have been previously defined on a Tag Menu Card) to a document. The lower part indicates which core interactions are combined

with and depending on the optional passage marking, the tag applies to a different scope. As for hyperlinks, this interaction technique supports flexible scopes. It can be used for tagging entire documents, passages within documents as well as collections of documents. Any time, the user can print a new version of a Tag Menu Card, in which previously defined tags are ordered and sized according to their frequency (tag cloud). Optionally, tags defined by all users or by members of the own learning group can be included.

Collecting all tags on a separate Tag Menu Card has the advantage that the user can immediately access a set of all tags. In addition, Tag Menu Cards support operations on the tag set (renaming etc.) which can then be automatically applied to the electronic representations of all documents and their subsequent printouts. Moreover, the approach supports co-located collaboration by allowing users to physically share cards. Finally, research shows that a key factor for the convergence of tags is that the system suggests frequent labels [31]. Yet, computer support cannot be assumed in a paper-only environment without a nearby display. In such a context, the Tag Menu Card fosters similar effects as the suggestion of frequently used tags: users will be inclined to re-using tags already entered on the card wherever possible, since the effort is lower than making a new tag entry.

Own and shared tags are displayed in the viewers for individual documents similarly to hyperlinks. Moreover, all tag concepts are automatically included in the ecological view as oval nodes (see Fig. 6.10 on p. 139). All documents, passages and collections of documents that are tagged with this concept are connected to it by an edge. This enables the user to quickly get an overview on all these contents.

7.3 Tagging with Buttons

In Section 5.2, we have described how buttons from a toolbar, which is printed on documents, can be used for defining the visibility of an annotation. CoScribe utilizes the same interaction technique and visualization for enabling users to tag individual annotations with semantic categories. The interaction technique is illustrated in Figs. 5.5 and 5.6 on p. 110f. This interaction is similar to applying tags using Tag Menu Cards with the difference that categories are predefined and that the buttons are printed on each document page, allowing for a quicker access.

CoScribe currently supports four semantic types, which were derived from the needs identified in our field studies. These are *Important*, *Question*, *To do* and *Correction*. Our approach also allows the user to select user-defined categories to be contained in the toolbar. However, if the user introduces new types, sharing and aggregation of annotations is more complicated due to the larger number of classes which moreover other users might not understand.

7.4 Tangible Tagging of Processes

All tagging strategies presented above are used for tagging *documents*. In order to structure and augment the collaborative *work process*, we propose a further concept: tagging processes using tangible objects. This supports users in jointly tagging the temporal phases of a co-located meeting. Such meetings often follow rather implicit, but predefined phases, for example starting by collecting all topics to discuss in a meeting, then discussing each topic in turn and finally planning tasks to perform until the next meeting.

Enabling the user to make these implicit phases more explicit by tagging processes has two main advantages. First, the system can temporally index and structure the users' activities (use of documents, annotations, links and tags) by attributing them to individual phases. This allows for a structured access. Second, the act of tagging itself is a meta-cognitive learning scaffold. Unexperienced learners might not know how to efficiently structure a meeting. To state two examples, they might not define clear goals for the meeting or they might be unaware of the importance of writing a protocol which contains main outcomes of the meeting. Offering specific types of phases which have proven to be important in meetings can stimulate users to follow these phases. Moreover, by stimulating users to explicitly specify the phase they are currently in, the system supports the meta-processes of negotiating and making joint decisions about the process structure.

The interaction technique aims at enabling users to specify the type of the process phase in which they are at the moment. Each time the group proceeds to a subsequent phase, the users indicate the new phase by manipulating a shared representation. For this reason, the system can temporally structure the users' activities (use of documents, annotations, links and tags) by attributing them to individual phases.

Main requirements for the interaction technique are first that it gives equal weight to all participants and enables all of them to specify types. This stimulates negotiation and joint group decisions. Second the technique should produce only minimal extraneous cognitive load in order to not disrupt the ongoing meeting. Moreover the currently selected phase should be clearly visible to all users with the goal to prevent wandering from the subject or even reverting to an unstructured meeting.

These requirements are met by a shared tangible object which can have different physical states and is located between the users on the table. A tangible is easy and intuitive to manipulate and its physical state indicates the current phase. We distinguish tangibles which stimulate by their affordances to follow a fix sequential structure from tangibles which suggest phases but do not suggest a specific sequence of them.

In order to illustrate the concept of tangible structuring of phases, we have developed two first prototypes of tangibles and an exemplary process model for learning group meetings.

Process Cube

The first prototype is a tangible cube of about 10 cm edge length, which is entirely covered with paper containing the Anoto pattern (Fig. 7.8). On each side of the cube, the paper covering has a different color, representing a minimum of two up to a maximum of six different phases in the collaborative process. Each time the group proceeds to a subsequent phase, the cube can be rotated to have the corresponding

Fig. 7.8 The Process Cube for collaborative tangible tagging of group processes

7.4 Tangible Tagging of Processes

Fig. 7.9 Exemplary model of semantic phases for structuring learning group meetings

side placed on top. The cube does not suggest a specific sequential structure of phases, since it can be arbitrarily rotated. As a shared object of all learners, the cube can be manipulated by all participants. The tangible three-dimensional cube provides for a very intuitive, shared representation and gives equal weight to all viewpoints of the participants located around it.

The system detects the cube's current state by sensing RFID tags which are located inside it. This automatically creates an index on a timeline in the ecological view, enabling the user to get an overview of the temporal structure of the meeting and to easily access all contents and activities of specific phases.

Documents can be linked to specific phases performing the association gesture on the corresponding side of the cube. Moreover, Digital Paper Post-its can be attached to the sides. This supports collaboratively collecting semantic items that are important and specific to a phase, e.g. topics to discuss in that meeting.

Based on existing models for structuring collaborative learning [61], we developed the following exemplary model that distinguishes six semantic phases of a learning group meeting (Fig. 7.9):

1. *Startup* (red): participants gather together and identify the topics of the meeting; Post-its represent topics.
2. *Explaining* (yellow): for the basic understanding of a topic, definitions and explanations are discussed; Post-its represent important properties of the topic.
3. *Elaboration* (orange): recall cues are collected (e.g. additional details, generated examples and images); Post-its represent recall cues for the topic.
4. *Abstraction* (gray): a topic is put in relation to existing knowledge and the entire learning content; Post-its represent links and levels of abstraction.
5. *Successfully accomplished* (green): contains the topics that have been successfully discussed and accomplished; Post-its represent topics.
6. *Ending/To Do* (blue): end phase of the meeting; agreement on tasks to perform individually after the meeting; Post-its represent to do items.

Fig. 7.10 The Process Knob for collaborative tangible tagging of group processes

Depending on the group processes to support, the cube can be easily modified to support other semantic types.

Process Knob

In contrast to the process cube, the second prototype exemplifies a tangible object that stimulates users to follow phases in a specific sequence. It is a rotary knob that can be rotated by 360 degrees. Around the knob, fields for the phases are arranged. Depending on the rotation, the needle points to one specific field and indicates one phase. The Process Knob is depicted in Figure 7.10.

Similarly to the cube, this tangible provides for intuitive manipulation of a clearly visible, shared representation. The circular arrangement affords rotating the knob a little further to make the needle point to the adjacent field each time the users proceed to the subsequent phase. Documents can be linked to specific phases as well, since the fields act as association areas for hyperlink gestures.

7.5 Evaluation and Discussion

In this chapter, we have presented four different interaction techniques for tagging documents and processes. These techniques complement each other, serving different purposes and having different properties. The choice of an appropriate technique depends both on the task and on the user's personal preferences. Table 7.2 gives a comparative overview of these techniques.

Digital Paper Bookmarks As shown in the overview table, Digital Paper Bookmarks can be used for tagging paper documents with user-defined tags and with

7.5 Evaluation and Discussion 163

Table 7.2 Comparison of the tagging techniques presented in this chapter

	Digital Paper Bookmarks	Tag Menu Card	Tag Buttons	Process Tagging
Documents / Process	D	D	D	P
Free Tagging	•	•	-	-
Category Tagging	•	-	•	•
Paper Documents	•	•	•	•
Digital Documents	-	•	•	•
Scope	Page	Document, Page, Passage	Annotation	Phase
Full visual feedback without computer	•	-	-	•

categories. While they are rather course-grained, applying only to entire pages, they offer the advantage of providing full visual feedback by their physical appearance.

We conducted an initial evaluation of Digital Paper Bookmarks as part of one of our user studies (see Section 5.4.2 for a detailed description of the study). We observed how nine participants used Digital Paper Bookmarks for indexing documents and for retrieving document passages. Feedback was gathered using a questionnaire and semi-structured interviews. The observations of how participants created and used bookmarks clearly indicate that the interaction technique is very easy to learn, easy to use and reliable. All participants readily understood the usage of bookmarks and created and modified bookmarks without assistance. Also in the questionnaire, bookmarks were rated as being very easy to create.[1] Concerning their use, bookmarks were judged to be helpful both for finding a specific page in the paper document[2] and for finding a page in the CoScribe viewer[3]. Several participants particularly valued the possibility to fade away from the given structure and to create instead an own structure of the document.

The visualization of the 3D paper stack with the bookmarks in the software viewer was considered to be very helpful for orientation within the document.[4] Several participants stated that this view provides a better awareness of the approximate current position within the document than the more traditional list that contains thumbnail images of the slides. Moreover, it takes less space than the list and provides a representation which resembles more closely the real paper stack. However, due to the 3D distortion of the visualization, the text on the bookmarks was less well readable if it was very small or written in a hurry. In this respect, the visualization of bookmarks directly on the slides is more powerful. These initial findings indicate a high usability of Digital Paper Bookmarks. They should be confirmed in a subse-

---

[1] $M = 4.7$ on a 5-point Likert scale, $SD = .5, N = 9$
[2] $M = 4.9, SD = .3, N = 9$
[3] $M = 4.4, SD = .5, N = 9$
[4] $M = 4.7, SD = .5, N = 9$

quent study, in which participants work with a larger collection of documents and use various tagging techniques.

Tag Menu Card As shown in the overview table above, the Tag Menu Card is used for tagging not only paper documents, but also digital representations of documents. It supports more flexible scopes than Digital Paper Bookmarks. A tag can be readily applied to an entire document, to a single page of a document or to another individually defined scope within a document. By introducing with the paper card the concept of a separate paper "tool", we provide physical instantiations of the tag sets, which remain inaccessible in previous paper-based systems. These instantiations stimulate the convergence of tags, operations on the entire tag set and sharing of tag sets.

Button Tagging We have gathered initial feedback on button tagging by deploying it in a field study. A total of 29 students used the technique during one of their regular lectures. Section 5.4.1 provides a more detailed description of the study. The results showed that the participants classified a substantial proportion of their annotations with a semantic category. 18.7 % of all annotations were tagged using the button tagging technique. This indicates that a few minutes of training are sufficient for learning how to use this technique.

The most frequently chosen category was 'Important' (12.5 %), followed by 'Question' (3.2 %), 'Correction' and 'To Do' (1.5 % each) (Fig. 7.11). Although the categories 'To do' and 'Correction' were used more rarely, several participants of the interviews explicitly stated that these types are nevertheless very important.

Several participants desired to define individual types like 'To revise', 'Definition' or 'Example'. From a designer's perspective, individual types are problematic because types are used to visualize shared annotations in their iconic form. With many different individual types defined by the users, other users would not be able to understand at a glance the meaning of shared annotations. An appropriate trade-off could consist of offering a set of some more common categories, of which the user would choose a subset to be printed on his or her personal printouts. In order to respond to the user feedback and to overcome the limitation of fix categories, we de-

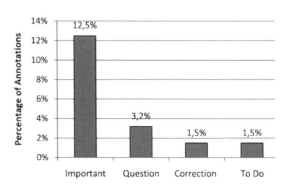

Fig. 7.11 Percentage of annotations that were classified with semantic categories ($N = 1983$)

veloped the technique of Tag Menu Cards for tagging with freely-chosen keywords and integrated this into the second version of our prototype.

Process Tagging The Process Cube and the Process Knob are two tangible tagging techniques that open up the design space of process tagging. Here, not contents of documents are tagged, but temporal phases in collaborative processes. On the one hand, tagging such phases allows for indexing artifacts that are used or created during the process. This eases future retrieval of artifacts. On the other hand, temporal process tagging can foster successfully structuring the process itself, as the tagging categories provide a scaffold. This scaffold supports participants of the meeting by suggesting them a good structure and by encouraging them to reflect and discuss about the structure. Subsequent research in the fields of design and computer science should identify further types of tangibles that meet the requirements. Research in social sciences should develop further process models and validate them.

Interaction Model Our interaction model of Pen-and-Paper User Interfaces prove beneficial as a basis for the design of the tagging techniques. It lead to simple and reliable interactions, to a varied and rich user experience as well as to a high degree of direct manipulation.

The interaction techniques are inspired by established practices of working with documents and let the user maintain them. For instance, users can attach an index sticker to a document page in order to tag this page. A small set of more formal interactions can be used to inform the system about the semantics of these informal artifacts. These interactions are based on a small set of recurrent core interactions. This leads to interaction that are easy to learn and to use and that are reliable even with the restricted feedback capabilities of Pen-and-Paper User Interfaces.

We have shown in this chapter that the novel interaction techniques draw upon the large variety of traditional paper-based practices. This includes writing with a pen on paper and manipulating the arrangement of physical objects. The techniques leverage tangible interaction by attaching physical stickers on a paper document and making pen-based associations between different documents. These varied interactions stand in contrast to paper interfaces that mimic the interaction of Graphical User Interfaces and are restricted to the interactions of writing, drawing symbols and clicking on paper surfaces.

The user can directly interact with documents using a digital pen and – for physical documents – his or her hands. This stands in contrast to the indirection caused by using a mouse. The interaction techniques leverage different types of paper-based tools, e.g. for defining collections of documents, for indexing document pages and for tagging documents. All these tools are not only instruments but also objects of interest, as they contain and represent first-class information. In contrast to typical tools in GUIs such as menus or toolbars, this double character lets the user perceive working directly on and with first class objects instead of interacting with tools. This results in a high degree of direct manipulation.

Chapter 8
Conclusions

Highly varied and efficient practices of interacting with physical documents have evolved over hundreds of years. Compared to these practices, the way we work today with documents on computers is limited in many respects. As a consequence, we are still far from becoming paperless. Most knowledge workers still make heavy use of printed documents, while they also use digital documents. This book has addressed Pen-and-Paper User Interfaces that bridge the gap between printed and digital documents. In this last chapter, we summarize the main outcome of this book and identify directions for emerging and future research.

8.1 Summary of this Book

While both traditional pen-and-paper practices and Graphical User Interfaces for document work are well-understood, it is a challenge to build a bridge between both worlds. In this book, we provided a survey of the state-of-the art, contributed a design approach with the first theoretical interaction model of Pen-and-Paper User Interfaces and presented a set of interaction techniques for annotating, linking and tagging documents.

State-of-the-Art Survey

This book has provided a comprehensive introduction to the field of Pen-and-Paper User Interfaces. A discussion of empirical studies on the use of paper from the literature has showed that paper has specific affordances, which are highly supportive for successful knowledge work. With respect to the readability of text, cutting-edge technology for electronic displays has almost caught up with the image quality of a good print on paper. However, paper still has a number of unique advantages, which are mostly due to its physical nature: paper documents make information tangible. They can be effectively navigated, annotated and organized in physical space.

In Chapter 2, we have reviewed prior research on user interfaces that integrate paper with computer support. The survey covered underlying technologies, technical frameworks and user interface concepts. It has shown that Pen-and-Paper Interfaces are technically mature, and have found their way into commercial applications. While the first generation of seminal systems addressed stationary setups at the knowledge worker's desk, a second generation focused on paper as a mobile medium, for instance, as augmented paper notebooks, and augmented printed documents. The survey has also shown that a growing body of research focuses on collaboration support and on integrating mobile use of pen and paper more closely with real-time feedback, for instance, by using augmented digital pens and leveraging mobile projection.

Interaction Model

We have presented an interaction model with the goal of contributing a theoretically-grounded approach for analysis and design of Pen-and-Paper User Interfaces. The overall model consists of a general design perspective, a model of interaction and a model of information. Empirical findings from the literature and from our own field studies have brought us to applying an ecological perspective. This perspective is inspired by distributed cognition and information ecologies. It accounts for both the tightly coupled interaction with physical and digital artifacts and for the prime relevance of collaboration that have been observed in the field. Based on the ecological perspective, we developed the first theoretical interaction model of Pen-and-Paper User Interfaces. By analytically separating the semantic level of interaction from the syntactic level of interaction, we have identified generic interaction primitives of paper-based user interfaces. Instead of mimicking interactions of Graphical User Interfaces, the interaction is geared to the varied traditional practices of using a pen and paper, and it leverages the rich interactions that are made possible by combining multiple interaction primitives. While we have shown that the model can be used to describe systems from previous work, the best proof of the applicability are the novel interaction techniques presented in this book. Based on the model, we presented highly versatile and intuitively usable interaction concepts that offer a rich user experience, since they are based on a set of tools which are made out of paper.

CoScribe: Interaction Techniques for Annotation, Linking and Tagging

In Chapters 4–7, we presented CoScribe, a concept and system for collaborative knowledge work. We introduced a coherent set of novel pen-based and tangible interaction techniques that, by their interplay, provide support for the key elements of collaborative knowledge work. These key elements consist of taking handwritten notes, of making handwritten annotations on documents, of creating and following cross-media hyperlinks between documents and of tagging documents and processes. CoScribe supports not only digital documents but enables the user also to

8.1 Summary of this Book

work with printed documents. Moreover, it enables various forms of collaboration around paper. Thereby, CoScribe closely combines multiple documents, users and processes in one integrated environment.

The design of CoScribe was guided by our theoretical interaction model. In the following, we discuss how the main aspects of this model – the ecological perspective and modular core interactions – influenced the design.

Ecological Perspective This holistic socio-technical approach directed our attention to the various relations that constitute information ecology. These are relations between multiple users, relations between users on the one hand and documents and tools they interact with on the other hand, relations between multiple documents, and relations between users and the practices of a given work setting. We now summarize by which means CoScribe supports each type of relation. These means are depicted in Fig. 8.1.

Relations between multiple users. CoScribe puts an emphasis on collaborative settings. Users can engage in co-located collaboration and remotely share documents and user-created content. Collaborative content can be accessed in the digital viewers or printed on paper. To the best of our knowledge, CoScribe is the first contribution that comprehensively addresses the challenges that are related to asynchronous sharing of handwritten content on paper. This comprises sharing own annotations and accessing shared annotations of other users. We presented a mechanism that enables users to share annotations directly on paper, which is is seamlessly integrated with paper-based annotation. This provides for flexible control of other persons' access to one's own annotations on various levels of visibility. Evaluation has shown that it is easy to learn and can be efficiently used after only some minutes of training. A current limitation is the small amount of feedback available in mo-

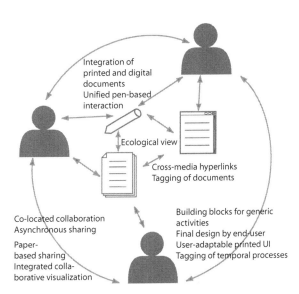

Fig. 8.1 Mapping of CoScribe's main concepts to the ecological perspective

bile settings where only a digital pen and paper is used. Novel generations of digital pens which include a display can overcome this limitation. Moreover, we have examined how shared handwritten annotations of many users can be visualized in an integrated manner and proposed a novel visualization that leverages a dynamic scaling approach for integrating annotations of all users in one single view. As shown by the results of a user study, this is particularly supportive for gaining an overview on shared annotations.

Relations between the user and documents/tools. CoScribe unifies the interaction with physical and digital artifacts. Regardless whether a document or a tool is available in printed form or visualized on a display, the same digital pen can be used to interact with it. As a consequence, the user does not have to switch between different input devices. One can even perform interactions that span paper and displays, for example linking a Web page with a printed document using one single pen gesture. This tight coupling is in contrast with most other paper-based applications and it outperforms augmented desk systems, which restrict the interaction to a small digitally augmented region.

Relations between multiple documents. Through its ecological perspective, CoScribe focuses on the relations that exist between multiple documents of a document collection. We have presented novel interaction techniques that allow the user to express these relations in order to integrate and structure collections of documents.

The first technique provides for creating, following and sharing hyperlinks between any combination of printed and digital documents. In contrast to previous work on cross-media hyperlinks, our technique supports the wide range of document types which is typical for information ecologies. This includes various formats of digital documents, their printed versions, Web pages and physical books. The technique puts strong emphasis on the associative character of hyperlinks, as a bidirectional hyperlink is created by an associative pen gesture that connects both link anchors. Links can be followed by tapping with the same pen on a link hot-spot either on paper, in the CoScribe viewer or in the Web browser. Finally, we have presented a collaborative ecological view that visualizes the relations within collections of documents. It provides a high-level overview on all hyperlinks and tags and provides a structured access to the documents of the collection. A user study has shown that hyperlinks between printed documents and Web pages lead to a significantly higher performance in information integration tasks with hybrid document collections than traditional references.

Moreover, we have contributed three novel techniques that offer support for structuring document collections with (predefined or freely-chosen) tags. The fourth technique tackles the question of how temporal processes can be tagged by collaborating users with tangible objects. The richness of the interaction techniques, which are based on our interaction model, becomes particularly noticeable here. The first technique leverages tangible stickers for tagging pages of physical documents. As we have seen, this is intuitive, provides full visual feedback on paper and affords quick access to a document page not only using a computer but also using the printed document instance. However, the tags are rather course-grained, since they always apply to entire pages. The second technique, which offers very flexible scopes, uses

8.1 Summary of this Book

a separate paper card for defining, applying and sharing freely-chosen tags. By introducing with the paper card the concept of a separate paper "tool", we provide physical instantiations of the tag sets, which remain inaccessible in previous paper-based systems. These instantiations allow for performing operations on the entire tag set, for sharing tag sets, and they stimulate the convergence of tags. The third technique aims at being particularly fast to be easily integrated into notetaking tasks. This is based on predefined categories that can be applied by tapping on specific buttons which are printed on paper documents. Finally, we have discussed the novel class of process tagging with tangible tools. We opened this research direction and proposed two first examples of interaction techniques. The variety of interaction metaphors for tagging stands in contrast to previous work on paper-based tagging which relies only on pen gestures.

Relations between the user and work practices. The theory of Information Ecologies postulates that technology should be designed in a way that leaves freedom of design to end-users. This allows them to adapt technology in order to symbiotically integrate it into the given practices of a local information ecology. In response to this postulation, CoScribe supports the generic activities of annotating, linking and tagging. As we have discussed, these can be used in very flexible ways for various purposes in different settings of knowledge work. As a consequence, CoScribe is not tailored to a specific purpose but offers a set of generic tools that end-users can flexibly combine and use according to his or her needs. This aspect has moreover directed our attention to the paper user interface, which in previous work is static and cannot be adapted by end-users. In contrast, CoScribe offers a user-adaptable paper interface.

Core Interactions Using the set of generic core interactions as a basis for the design of CoScribe's interaction techniques has proved beneficial. It leads to interactions that are easy to learn, easy to use and reliable, it leads to a varied and rich user experience, and it leads to a high degree of direct manipulation.

The interaction techniques allow the user to maintain the existing practices of working with a pen and printed documents. For instance, users can make handwritten annotations at any position within documents and create handwritten references and tags. This accounts for the highly individual practices observed in contextual inquiries. A small set of more formal interactions can be used to inform the system about the semantics of these informal artifacts. These interactions are based on a set of recurrent core interactions. Each core interaction is easy to learn and to use and can be performed reliably even with the restricted feedback capabilities of most current digital pens.

We have shown that the novel interaction techniques draw upon the richness of traditional paper-based practices. This includes writing with a pen on paper, manipulating the orientation of physical objects and creating physical arrangements of multiple documents, e.g. in paper folders. Further techniques leverage tangible interaction by attaching physical stickers on a paper document and making pen-based associations between different documents. These varied interactions stand in contrast to paper interfaces that mimic the interaction of Graphical User Interfaces and that are restricted to the interactions of writing, drawing symbols and clicking and to using one single sheet of paper.

The user can directly interact with documents using a digital pen and – for physical documents – his or her hands. This is in contrast to the indirection caused by using a mouse. The interaction techniques leverage different types of paper-based tools, e.g. for defining collections of documents, for indexing document pages and for tagging documents. All these tools are not only instruments but also objects of interest, as they contain and represent first-class information. In contrast to typical tools in GUIs such as menus or toolbars, this double character lets users perceive working directly on and with first-class objects instead of interacting with instruments. This results in a high degree of direct manipulation.

8.2 Directions of Future Research

To conclude this book, we outline promising future directions of research on Pen-and-Paper User Interfaces.

Enhancing Real-time Output on Paper

Most previous work has focused on using a digital pen and paper for data input; most often output was provided not on paper, but rather on a separate computer screen. Future work will aim at enhancing real-time output that is provided at the place where input is happening. This comprises enhanced pens that provide feedback on the pen, pens that overlay printed information with projected digital information, and pens that feature a built-in inkjet printer. While this brings traditional paper closer to electronic displays, a complementary direction consists of bringing computer displays closer to paper. We will discuss both directions in turn. In what follows, we will discuss both directions.

Livescribe has indicated the direction that future pens are likely to follow further: include more powerful processors which execute applications directly on the pen and provide visual and auditory feedback in real-time. Current Livescribe pens provide auditory and visual output, but they are restricted by a small display and a missing wireless data connection. Future pens are likely to feature a larger display – eventually the entire surface of the pen might become a display – and more varied haptic feedback.

In order to provide real-time digital output directly on paper future pens are likely to feature a mobile projector, as suggested by Song et al. [138]. Alternatively, a second device, different from the pen, will be used for projecting contents. While in non-mobile settings, the projector will be integrated into a mouse [139], a desk lamp [14] or another common device, it is most likely that in mobile settings, the device of choice will be a mobile phone. Mobile tracking-projection solutions are promising for transforming any paper document into an interactive surface. The

8.2 Directions of Future Research

increasing processing power of mobile phones and the advent of very small mobile projection units let us expect that in the near future, every smartphone comprises the components that are required for creating paper-digital interactive surfaces on-the-fly. Future work should examine more deeply interaction techniques for mobile Pen-and-Paper User Interfaces that feature a camera-projection unit. Promising first steps are presented in [77, 78].

Projection creates ephemeral visual output. The paper document itself remains unchanged. We envision future pens to feature a small, built-in inkjet printer. This can be used to leave permanent marks on paper. Such devices will allow us to re-define the boundary between ephemeral digital output and output that permanently alters the printed user interface. A first step towards this direction is presented in [81].

Flexible OLED displays and electronic paper [16] are highly promising trends. These novel technologies allow for displays that are very thin and deformable, that feature a very high contrast, are well readable even in difficult light conditions and consume only little energy. However, the update rate of electronic paper is currently too low for interactive applications, and current deformable OLED displays suffer from a short life span. Future OLED and electronic paper displays will have the potential to replace traditional paper to a much higher degree than current display technologies. Our vision consists of an environment, in which the user disposes of a large number of thin and malleable displays, which can be manipulated very much like traditional paper. In contrast to current displays, multiple display surfaces allow the user to lay out pages in the physical space, to compare pages and to arrange them in flexible ways. As pointed out in [40], pages of digital documents would be temporarily bound to displays. In this context, it is an important question to what extent and how existing Pen-and-Paper interaction techniques can be transferred to such an environment. This should be examined in future work. We assume that annotation-based and printed-button-based techniques can be transferred in a straightforward manner. Moreover, all interactions that rely on associations between multiple page surfaces (e.g. linking, Menu Card tagging) apply to a multi-page display environment as well. Hence, a large proportion of the techniques presented in this book generalize not only to paper but also to future paper-like displays.

A significant difference between paper and a paper-like display is that information on paper is fixed while it is only temporarily bound to a display. This provides for more flexibility of use. However, if the user disposes only of a restricted number of flexible displays, he or she is forced to successive use of one display for different documents or pages. This challenges tangible paper-based techniques. For instance Digital Paper Bookmarks would have to automatically appear and disappear at the edges of the paper display. For all kinds of physical indexing, it would be highly supportive to develop flexible displays that are able to change their physical shape. In this context, actuation based on shape memory alloys is highly promising (cf. [17, 121]). A further challenge are filing interactions, such as placing a document into a folder. Traditional paper would remain in the folder and thereby physically indicate its contents, the number of documents filed, etc. On the contrary, if the flexible display is reused and does not remain within the folder, this physical information gets lost. Future research should therefore not only address the question of how a set of flexible displays can be quickly bound to a set of documents but also the issue of how interaction techniques and tangible tools can compensate for the physical information that gets lost when removing the unambiguous mapping between a document page and its physical carrier surface.

Towards Fully Mobile Pen-and-Paper Interfaces

Most prior applications either cannot be used at all in mobile settings or only parts of their functionality are available in a mobile setting. This is due to the fact that most of the more complex functionality requires that data be made available to a back-end system. While today's digital pens can be used in a fully mobile mode when data is kept on the pen, sending data in real-time to a back-end system requires additional hardware. In this case, current pens require another mobile device they can stream pen data to via a Bluetooth connection.

Only a very limited number of previous systems has used this setup (most notably the EdFest system [134, p. 153 ff.]). In the future, we expect more applications that couple paper and a digital pen with a mobile phone. The advantage of this device federation is that mobile phones are widely deployed and fit perfectly to the mobile setup. The mobile phone can act as a communication hub between the pen and the network. It can run applications, and it can provide feedback on a display or via auditory and haptic channels. Hence, the coupling of pen, paper and a mobile phone requires only standard hardware and results in a powerful device federation. Future work should examine how the user interface can be repartitioned between pen-and-paper and the mobile phone. If the same pen can be used to interact on paper and on the display of the mobile phone, paper and display merge in a way that is similar to the interaction techniques presented in this book. The difference is that in contrast to our techniques, not smaller paper is placed on a larger display, but a small display is placed on paper, which creates new challenges.

Support of Further Manipulations of Paper

In our work, we deliberately decided not to augment paper by electronic components nor to use an extensive solution for tracking paper in physical space. This allowed us to maintain paper as an inexpensive medium, on which digital contents can be easily printed, and which can be used virtually everywhere. A drawback of this approach is that some interactions with paper cannot be captured. Our core interactions rely on writing with the pen on paper and on changing the physical location of paper pages. Other interactions, such as folding a sheet of paper, are not comprised.

Emerging novel tracking technologies will likely resolve this trade-off. Recently, inexpensive depth cameras, such as the Microsoft Kinect[1], have come to the mar-

---

[1] http://www.xbox.com/kinect

ket that provide a high-resolution depth image of a physical scenery with a high framerate. The depth image allows to easily capture the deformation of a sheet of paper. Also the use of flexible electronic paper and OLED displays – as discussed above – would make our premise obsolete, since the user would not print contents on paper but would use an electronic device which can be enhanced with sensors for additional core interactions. Future work should examine further core interactions of Pen-and-Paper User Interfaces. This concerns manipulations that deform sheets of paper, covering interactions such as bending [128, 163], folding [70, 71], rolling [70, 56], and tearing a sheet in two.

Improving Large-scale Collaboration with Paper

A further challenge for future Pen-and-Paper Interfaces is improving collaboration support. Most current applications focus on a single user. Sharing of handwritten contents with other users is only supported by a very small number of systems. In particular, it is still not fully understood how to process, integrate and visualize paper-based contents that are created by a very large community of users. This point is related to the automatic interpretation of contents. Almost all current systems interpret pen-and-paper interactions only to a very limited extent. A small set of gestures might be interpreted (e.g. for creating hyperlinks), but the semantics of the remaining handwritten contents is not recognized. The systems typically display only a facsimile of the handwritten contents, sometimes performing handwriting recognition in the background to allow for full-text search. While this is sufficient for a single user or for small groups, it is clearly not adequate when it comes to integrating contents of a large number of users. It will be interesting to see how contents that are created on pen-and-paper can be integrated with tagging platforms, blogs and social networks, known as the Web 2.0. In the context of our interaction techniques, this could for example comprise automatically aggregating shared annotations, links and tags and recommending relevant annotations, documents and passages.

Standards and Interoperability

The field is currently characterized by a large number of research prototypes and a some commercial applications, each of them forming an individual island solution. As pointed out by Signer et al. [136], an adoption by the mass market requires more standards and a better interoperability of solutions. This comprises interface standards that allow application developers to abstract from the specific hardware solution for pen input. Moreover, there is a need for a common representation of digital ink data. While with InkML[2] a W3C standard exists, this standard is very complex and therefore has not become widely accepted yet.

---

[2] http://www.w3.org/TR/InkML

Another challenge is related to authoring and publishing. On the one hand, applications should enable end-users to easily create their own pen-enabled documents and paper interfaces. While for a long time, this was possible only with research prototypes, first commercial applications targeting at end-users are now issued on the market. On the other hand, current pen-enabled print products are closed solutions. Only if a user disposes of the specific application that belongs to the product, he or she can interact with the print product. However, a mass market of pen-enabled print products requires interoperability between different applications. In this context, a service-based approach seems promising [136, 36].

Better Understanding of Long-term Usage

Finally, we have seen that, even though a large number of Pen-and-Paper Interfaces have been developed over the past years, there is only very little research that examines *how* people use these interfaces. Only a very small number of studies examine how Pen-and-Paper Interfaces are used over a longer period of time and how they get integrated into existing information ecologies [89, 149, 147]. There is a clear need for a better understanding of how interfaces are used over longer periods of time and how they affect our work practices. This will take us another step further on the challenging and exciting journey towards closing the gap between printed and digital documents.

References

1. Adler, A., Gujar, A., Harrison, B.L., O'Hara, K., Sellen, A.: A diary study of work-related reading: design implications for digital reading devices. In: CHI '98: Proceedings of the SIGCHI conference on Human factors in computing systems, pp. 241–248. ACM Press/Addison-Wesley Publishing Co., New York, NY, USA (1998)
2. Adler, M.J., Doren, C.V.: How to Read a Book. Revised Edition. Simon and Schuster (1972)
3. Aitenbichler, E., Kangasharju, J., Mühlhäuser, M.: MundoCore: A Light-weight Infrastructure for Pervasive Computing. Pervasive and Mobile Computing (2007)
4. Anoto: Digital pen and paper technology. Anoto AB. http://www.anoto.com
5. Arai, T., Aust, D., Hudson, S.E.: Paperlink: a technique for hyperlinking from real paper to electronic content. In: CHI '97: Proceedings of the SIGCHI conference on Human factors in computing systems, pp. 327–334. ACM, New York, NY, USA (1997)
6. Back, M., Cohen, J., Gold, R., Harrison, S., Minneman, S.: Listen reader: an electronically augmented paper-based book. In: CHI '01: Proceedings of the SIGCHI conference on Human factors in computing systems, pp. 23–29. ACM, New York, NY, USA (2001)
7. Bandyopadhyay, D., Raskar, R., Fuchs, H.: Dynamic shader lamps: Painting on movable objects. In: International Symposium on Augmented Reality, p. 207. IEEE Computer Society, Los Alamitos, CA, USA (2001)
8. Beaudouin-Lafon, M.: Instrumental interaction: an interaction model for designing post-wimp user interfaces. In: CHI '00: Proceedings of the SIGCHI conference on Human factors in computing systems, pp. 446–453. ACM, New York, NY, USA (2000)
9. Block, F., Haller, M., Gellersen, H., Gutwin, C., Billinghurst, M.: Voodoosketch: Physical interface palettes and sketched controls alongside augmented work surfaces. In: Adjunct. Proceedings of UbiComp 07, pp. 789–793 (2007)
10. Brandl, P., Haller, M., Hurnaus, M., Lugmayr, V., Oberngruber, J., Oster, C., Schafleitner, C., Billinghurst, M.: An adaptable rear-projection screen using digital pens and hand gestures. In: 17th International Conference on Artificial Reality and Telexistence, pp. 49–54 (2007)
11. Buechley, L., Hendrix, S., Eisenberg, M.: Paints, paper, and programs: First steps toward the computational sketchbook. In: Proceedings of the Third International Conference on Tangible and Embedded Interaction (TEI'09), pp. 9–12 (2009)
12. Bush, V.: As we may think. The Atlantic Monthly pp. 101–108 (1945)
13. Cepiprint - Association of European Publication Paper Producers: Demand and supply statistics 1990-2010. http://www.cepiprint.com/img_clipart/DS_2011_final.pdf (2010)
14. Chan, L.W., Wu, H.T., Kao, H.S., Ko, J.C., Lin, H.R., Chen, M.Y., Hsu, J., Hung, Y.P.: Enabling beyond-surface interactions for interactive surface with an invisible projection. In: Proceedings of the 23rd annual ACM symposium on User interface software and technology, UIST '10, pp. 263–272. ACM, New York, NY, USA (2010)
15. Chandler, P., Sweller, J.: Cognitive load theory and the format of instruction. Cognition and Instruction **8**(4), 293–332 (1991)

16. Co, E., Pashenkov, N.: Emerging display technologies for organic user interfaces. Commun. ACM **51**(6), 45–47 (2008)
17. Coelho, M., Hall, L., Berzowska, J., Maes, P.: Pulp-based computing: A framework for building computers out of paper. In: The 9th International Conference on Ubiquitous Computing (Ubicomp 2007) (2007)
18. Conroy, K.M., Levin, D., Guimbretière, F.: Proofrite: A paper-augmented word processor. Tech. rep., University of Maryland (2004)
19. Coutaz, J., Nigay, L., Salber, D., Blandford, A., May, J., Young, R.M.: Four easy pieces for assessing the usability of multimodal interaction: The care properties. In: Proceedings of INTERACT'95 (1995)
20. Cowan, L.G., Weibel, N., Pina, L.R., Hollan, J.D., Griswold, W.G.: Ubiquitous sketching for social media. In: Proc. of MobileHCI 2011. ACM Press (2011)
21. van Dam, A.: Post-wimp user interfaces. Communications of the ACM **40**(2), 63–67 (1997)
22. Decker, C., Beigl, M., Eames, A., Kubach, U.: Digiclip: Activating physical documents. In: ICDCSW '04: Proceedings of the 24th International Conference on Distributed Computing Systems Workshops - W7: EC (ICDCSW'04), pp. 388–393. IEEE Computer Society, Washington, DC, USA (2004)
23. Dillon, A.: Reading from paper versus screens: a critical review of the empirical literature. Ergonomics **35**(10), 1297–1326 (1992)
24. Doermann, D.: The indexing and retrieval of document images: a survey. Comput. Vis. Image Underst. **70**, 287–298 (1998)
25. Drucker, P.F.: The concept of the corporation. New American Library of World Literature (1964)
26. Erol, B., Antúnez, E., Hull, J.J.: Hotpaper: multimedia interaction with paper using mobile phones. In: Proceeding of the 16th ACM international conference on Multimedia, MM '08, pp. 399–408. ACM, New York, NY, USA (2008)
27. Everitt, K., Morris, M.R., Brush, A.B., Wilson, A.: Docudesk: An interactive surface for creating and rehydrating many-to-many linkages among paper and digital documents. In: Proceedings of IEEE Tabletops and Interactive Surfaces 2008, pp. 25–28 (2008)
28. Fiala, M.: Artag, an improved marker system based on artoolkit. Tech. Rep. NRC 47166, National Research Council Canada (2004)
29. Fishkin, K.P.: A taxonomy for and analysis of tangible interfaces. Personal Ubiquitous Comput. **8**(5), 347–358 (2004)
30. Foley, J.D., van Dam, A., Feiner, S.K., Hughes, J.F.: Computer graphics: principles and practice (2nd ed.). Addison-Wesley Longman Publishing Co., Inc., Boston, MA, USA (1990)
31. Golder, S., Huberman, B.A.: Usage patterns of collaborative tagging systems. Journal of Information Science **32**(2), 198–208 (2006)
32. Guimbretière, F.: Paper augmented digital documents. In: UIST '03: Proceedings of the 16th annual ACM symposium on User interface software and technology, pp. 51–60. ACM Press, New York, NY, USA (2003)
33. Haller, M., Leithinger, D., Leitner, J., Seifried, T., Brandl, P., Zauner, J., Billinghurst, M.: The shared design space. In: ACM SIGGRAPH 2006 Emerging technologies, SIGGRAPH '06. ACM, New York, NY, USA (2006)
34. Hartmann, B., Morris, M.R., Benko, H., Wilson, A.D.: Pictionaire: supporting collaborative design work by integrating physical and digital artifacts. In: CSCW '10: Proceedings of the 2010 ACM conference on Computer supported cooperative work, pp. 421–424. ACM, New York, NY, USA (2010)
35. Heiner, J.M., Hudson, S.E., Tanaka, K.: Linking and messaging from real paper in the paper pda. In: UIST '99: Proceedings of the 12th annual ACM symposium on User interface software and technology, pp. 179–186. ACM Press, New York, NY, USA (1999)
36. Heinrichs, F., Steimle, J., Schreiber, D., Mühlhäuser, M.: Letras: an architecture and framework for ubiquitous pen-and-paper interaction. In: Proceedings of the 2nd ACM SIGCHI symposium on Engineering interactive computing systems, EICS '10, pp. 193–198. ACM, New York, NY, USA (2010)

37. Hinckley, K., Ramos, G., Guimbretiere, F., Baudisch, P., Smith, M.: Stitching: pen gestures that span multiple displays. In: AVI '04: Proceedings of the working conference on Advanced visual interfaces, pp. 23–31. ACM Press, New York, NY, USA (2004)
38. Hofer, R.L., Kunz, A.: Digisketch: taming anoto technology on lcds. In: Proceedings of the 2nd ACM SIGCHI symposium on Engineering interactive computing systems, EICS '10, pp. 103–108. ACM, New York, NY, USA (2010)
39. Hollan, J., Hutchins, E., Kirsh, D.: Distributed cognition: Toward a new foundation for human-computer interaction research. ACM Transactions on Human-Computer Interaction **7**(2), 174–196 (2000)
40. Holman, D., Vertegaal, R., Altosaar, M., Troje, N., Johns, D.: Paper windows: interaction techniques for digital paper. In: CHI '05: Proceedings of the SIGCHI conference on Human factors in computing systems, pp. 591–599. ACM Press, New York, NY, USA (2005)
41. Hornecker, E., Buur, J.: Getting a grip on tangible interaction: a framework on physical space and social interaction. In: CHI '06: Proceedings of the SIGCHI conference on Human Factors in computing systems, pp. 437–446. ACM, New York, NY, USA (2006)
42. Hutchins, E.: Cognition in the Wild. MIT Press (1995)
43. Ishii, H., Ullmer, B.: Tangible bits: towards seamless interfaces between people, bits and atoms. In: CHI '97: Proceedings of the SIGCHI conference on Human factors in computing systems, pp. 234–241. ACM, New York, NY, USA (1997)
44. ISO International Organization for Standardization: ISO/IEC Standard 15420:2000 (Bar code symbology specification - EAN/UPC) (2000)
45. ISO International Organization for Standardization: ISO/IEC Standard 16022:2000 (International symbology specification - Data Matrix) (2000)
46. ISO International Organization for Standardization: ISO/IEC Standard 18004:2006 (QR Code 2005 bar code symbology symbology specification) (2006)
47. Izadi, S., Hodges, S., Taylor, S., Rosenfeld, D., Villar, N., Butler, A., Westhues, J.: Going beyond the display: a surface technology with an electronically switchable diffuser. In: UIST '08: Proceedings of the 21st annual ACM symposium on User interface software and technology, pp. 269–278. ACM, New York, NY, USA (2008)
48. Jacob, R.J., Girouard, A., Hirshfield, L.M., Horn, M.S., Shaer, O., Solovey, E.T., Zigelbaum, J.: Reality-based interaction: a framework for post-wimp interfaces. In: CHI '08: Proceeding of the twenty-sixth annual SIGCHI conference on Human factors in computing systems, pp. 201–210. ACM, New York, NY, USA (2008)
49. Jervis, M., Masoodian, M.: Digital management and retrieval of physical documents. In: Proceedings of the Third International Conference on Tangible and Embedded Interaction (TEI'09), pp. 47–54 (2009)
50. Jiang, H., Yeh, R.B., Winograd, T., Shi, Y.: Digipost: Writing on post-its with digital pen to support collaborative editing tasks on tabletop displays. In: ACM symposium on User interface software and technology (UIST'07) Posters (2007)
51. Jonassen, D.H., Beissner, K., Yacci, M.: Structural Knowledge: Techniques for Representing, Conveying, and Acquiring Structural Knowledge. Lawrence Erlbaum Associates, Hillsdale (1993)
52. Kafka, C.: Dataglyphs bridge paper and digital worlds. Docu World pp. 66–67 (1998)
53. Kaltenbrunner, M., Bencina, R.: reactivision: a computer-vision framework for table-based tangible interaction. In: Proceedings of the 1st international conference on Tangible and embedded interaction, TEI '07, pp. 69–74. ACM, New York, NY, USA (2007)
54. Kantola, V., Kulovesi, J., Lahti, L., Lin, R., Zavodchikova, M., Coatanea, E.: Printed electronics, now and future. In: Y. Neuvo, S. Ylönen (eds.) Bit Bang - Rays to the Future. Helsinki University Print (2009)
55. Kaplan, F., Do-Lenh, S., Dillenbourg, P.: Docklamp: a portable projector-camera system. In: 2nd IEEE TABLETOP Workshop (2007)
56. Khalilbeigi, M., Lissermann, R., Mühlhäuser, M., Steimle, J.: Xpaaand: Interaction techniques for rollable displays. In: CHI '11: Proceedings of the 29th international conference on Human factors in computing systems. ACM Press (2011)

57. Khalilbeigi, M., Steimle, J., Mühlhäuser, M.: Interaction techniques for hybrid piles of documents on interactive tabletops. In: CHI 2010 Extended Abstracts on Human Factors in Computing Systems, pp. 3943–3948 (2010)
58. Kiewra, K.A.: Notetaking and review: The research and its implications. Instructional Science 16(3), 233–249 (1987)
59. Kiewra, K.A.: A review of note-taking: The encoding-storage paradigm and beyond. Educational Psychology Review 1(2), 147–172 (1989)
60. Kim, J., Seitz, S.M., Agrawala, M.: Video-based document tracking: unifying your physical and electronic desktops. In: UIST '04: Proceedings of the 17th annual ACM symposium on User interface software and technology, pp. 99–107. ACM, New York, NY, USA (2004)
61. King, A.: Scripting collaborative learning processes: A cognitive perspective. In: F. Fischer, I. Kollar, H. Mandl, J.M. Haake (eds.) Computer-Supported Collaborative Learning, chap. 2, pp. 13–37. Springer US (2007)
62. Klemmer, S.R., Graham, J., Wolff, G.J., Landay, J.A.: Books with voices: paper transcripts as a physical interface to oral histories. In: CHI '03: Proceedings of the SIGCHI conference on Human factors in computing systems, pp. 89–96. ACM Press, New York, NY, USA (2003)
63. Klemmer, S.R., Newman, M.W., Farrell, R., Bilezikjian, M., Landay, J.A.: The designers' outpost: a tangible interface for collaborative web site design. In: UIST '01: Proceedings of the 14th annual ACM symposium on User interface software and technology, pp. 1–10. ACM Press, New York, NY, USA (2001)
64. Koike, H., Sato, Y., Kobayashi, Y.: Integrating paper and digital information on enhanceddesk: a method for realtime finger tracking on an augmented desk system. ACM Trans. Comput.-Hum. Interact. 8, 307–322 (2001)
65. Koile, K., Chevalier, K., Rbeiz, M., Rogal, A., Singer, D., Sorensen, J., Smith, A., Tay, K.S., Wu, K.: Supporting feedback and assessment of digital ink answers to in-class exercises. In: Nineteenth Conference on Innovative Applications of AI, pp. 22–29 (2007)
66. Koleva, B., Benford, S., Ng, K.H., Rodden, T.: A framework for tangible user interfaces. In: Physical Interaction Workshop on Real World User Interfaces in conjunction with Mobile HCI Conference 2003 (2003)
67. Krasner, G.E., Pope, S.T.: A description of the model-view-controller user interface paradigm in the smalltalk-80 system. Journal of Object Oriented Programming 1(3), 26–49 (1988)
68. Lai, W.C., Chao, P.Y., Chen, G.D.: The interactive multimedia textbook: Using a digital pen to support learning for computer programming. In: IEEE International Conference on Advanced Learning Technologies (ICALT'07), pp. 742–746 (2007)
69. Lee, B., Maldonado, H., Klemmer, S.R., Kim, I., Hilfinger-Pardo, P.: Longitudinal studies of augmented notebook usage informing the design of sharing mechanisms. Tech. Rep. CSTR 2006-11 09/29/06, Stanford University (2006)
70. Lee, J.C., Hudson, S.E., Tse, E.: Foldable interactive displays. In: UIST '08: Proceedings of the 21st annual ACM symposium on User interface software and technology, pp. 287–290. ACM, New York, NY, USA (2008)
71. Lee, S.S., Kim, S., Jin, B., Choi, E., Kim, B., Jia, X., Kim, D., Lee, K.p.: How users manipulate deformable displays as input devices. In: CHI '10: Proceedings of the 28th international conference on Human factors in computing systems, pp. 1647–1656. ACM, New York, NY, USA (2010)
72. Levenshtein, V.I.: Binary codes capable of correcting deletions, insertions, and reversals. Soviet Physics Doklady 10, 707–710 (1966)
73. Li, Y., Hinckley, K., Guan, Z., Landay, J.A.: Experimental analysis of mode switching techniques in pen-based user interfaces. In: CHI '05: Proceedings of the SIGCHI conference on Human factors in computing systems, pp. 461–470. ACM, New York, NY, USA (2005)
74. Liao, C., Guimbretière, F., Anderson, R., Linnell, N., Prince, C., Razmov, V.: Papercp: Exploring the integration of physical and digital affordances for active learning. In: Human-Computer Interaction INTERACT 2007, pp. 15–28 (2007)
75. Liao, C., Guimbretière, F., Hinckley, K., Hollan, J.: Papiercraft: A gesture-based command system for interactive paper. ACM Transactions on Computer-Human Interaction 14(4), 1–27 (2008)

References

76. Liao, C., Guimbretière, F., Loeckenhoff, C.E.: Pen-top feedback for paper-based interfaces. In: UIST '06: Proceedings of the 19th annual ACM symposium on User interface software and technology, pp. 201–210. ACM Press, New York, NY, USA (2006)
77. Liao, C., Liu, Q., Liew, B., Wilcox, L.: Pacer: fine-grained interactive paper via camera-touch hybrid gestures on a cell phone. In: Proceedings of the 28th international conference on Human factors in computing systems, CHI '10, pp. 2441–2450. ACM, New York, NY, USA (2010)
78. Liao, C., Tang, H., Liu, Q., Chiu, P., Chen, F.: Fact: fine-grained cross-media interaction with documents via a portable hybrid paper-laptop interface. In: Proceedings of the international conference on Multimedia, MM '10, pp. 361–370. ACM, New York, NY, USA (2010)
79. Liu, Q., Yano, H., Kimber, D., Liao, C., Wilcox, L.: High accuracy and language independent document retrieval with a fast invariant transform. In: Proceedings of the 2009 IEEE international conference on Multimedia and Expo, ICME'09, pp. 386–389. IEEE Press, Piscataway, NJ, USA (2009)
80. Liwicki, M., Eisha, H.M.A., Dengel, A.: Improving handwriting recognition by the use of semantic information. In: Proceedings of the 9th IAPR International Workshop on Document Analysis Systems, DAS '10, pp. 441–446. ACM, New York, NY, USA (2010)
81. Liwicki, M., Uchida, S., Iwamura, M., Omachi, S., Kise, K.: Data-embedding pen: augmenting ink strokes with meta-information. In: Proceedings of the 9th IAPR International Workshop on Document Analysis Systems, DAS '10, pp. 43–52. ACM, New York, NY, USA (2010)
82. Ljungstrand, P., Redström, J., Holmquist, L.E.: Webstickers: using physical tokens to access, manage and share bookmarks to the web. In: DARE '00: Proceedings of DARE 2000 on Designing augmented reality environments, pp. 23–31. ACM Press, New York, NY, USA (2000)
83. Lowe, D.G.: Distinctive image features from scale-invariant keypoints. Int. J. Comput. Vision **60**, 91–110 (2004)
84. Luff, P., Heath, C.: Mobility in collaboration. In: Proceedings of the 1998 ACM conference on Computer supported cooperative work, CSCW '98, pp. 305–314. ACM, New York, NY, USA (1998)
85. Mackay, W., Pagani, D.: Video mosaic: laying out time in a physical space. In: Proceedings of the second ACM international conference on Multimedia, MULTIMEDIA '94, pp. 165–172. ACM, New York, NY, USA (1994)
86. Mackay, W., Velay, G., Carter, K., Ma, C., Pagani, D.: Augmenting reality: adding computational dimensions to paper. In: Communications of the ACM, vol. 36, pp. 96–97. ACM, New York, NY, USA (1993)
87. MacKay, W.E.: Is paper safer? the role of paper flight strips in air traffic control. ACM Transactions on Computer-Human Interaction **6**(4), 311–340 (1999)
88. Mackay, W.E., Pothier, G., Letondal, C., Bøegh, K., Sørensen, H.E.: The missing link: augmenting biology laboratory notebooks. In: UIST '02: Proceedings of the 15th annual ACM symposium on User interface software and technology, pp. 41–50. ACM Press, New York, NY, USA (2002)
89. Maldonaldo, H., Lee, B., Klemmer, S.R., Pea, R.D.: Patterns of collaboration in design courses: Team dynamics affect technology appropriation, artifact creation, and course performance. In: Proc. of Computer Supported Collaborative Learning (CSCL) 2007 (2007)
90. Malone, T.W.: How do people organize their desks?: Implications for the design of office information systems. ACM Trans. Inf. Syst. **1**(1), 99–112 (1983)
91. Marshall, C.C.: Annotation: from paper books to the digital library. In: DL '97: Proceedings of the second ACM international conference on Digital libraries, pp. 131–140. ACM Press, New York, NY, USA (1997)
92. Marshall, C.C.: Toward an ecology of hypertext annotation. In: HYPERTEXT '98: Proceedings of the ninth ACM conference on Hypertext and hypermedia : links, objects, time and space—structure in hypermedia systems, pp. 40–49. ACM Press, New York, NY, USA (1998)

93. Marshall, J., Pridmore, T., Pound, M., Benford, S., Koleva, B.: Pressing the flesh: Sensing multiple touch and finger pressure on arbitrary surfaces. In: Proceedings of the 6th International Conference on Pervasive Computing, Pervasive '08, pp. 38–55. Springer-Verlag, Berlin, Heidelberg (2008)
94. McGee, D., Huang, X., Barthelmess, P., Cohen, P.: The neteyes collaborative, augmented reality, digital paper system. In: Proc. of IEEE Symposium on 3D User Interfaces 2008 (2008)
95. Mistry, P., Maes, P.: Intelligent sticky notes that can be searched, located and can send reminders and messages. In: International Conference on Intelligent User Interfaces (IUI'08), pp. 425–426 (2008)
96. Mistry, P., Maes, P., Chang, L.: Wuw - wear ur world: a wearable gestural interface. In: Proceedings of the 27th international conference extended abstracts on Human factors in computing systems, CHI '09, pp. 4111–4116. ACM, New York, NY, USA (2009)
97. Miura, M., Kunifuji, S., Sakamoto, Y.: Airtransnote: An instant note sharing and reproducing system to support students learning. In: IEEE International Conference on Advanced Learning Technologies (ICALT), pp. 175–179 (2007)
98. Müller-Tomfelde, C., Fjeld, M.: Introduction: A short history of tabletop research, technologies, and products. In: C. Müller-Tomfelde (ed.) Tabletops - Horizontal Interactive Displays. Springer (2010)
99. Morris, M.R., Brush, A.B., Meyers, B.R.: Reading revisited: Evaluating the usability of digital display surfaces for active reading tasks. In: International Workshop on Horizontal Interactive Human-Computer Systems (2007)
100. Nardi, B., O'Day, V.: Information Ecologies: Using Technology with Heart, chap. Chapter Four: Information Ecologies. MIT Press (1999)
101. Nelson, L., Ichimura, S., Pedersen, E.R., Adams, L.: Palette: a paper interface for giving presentations. In: CHI '99: Proceedings of the SIGCHI conference on Human factors in computing systems, pp. 354–361. ACM Press, New York, NY, USA (1999)
102. Nelson, T.H.: Complex information processing: a file structure for the complex, the changing and the indeterminate. In: Proceedings of the ACM 20th national conference, pp. 84–100. ACM, New York, NY, USA (1965)
103. Nielsen, J.: ipad and kindle reading speeds. Jakob Nielsen's Alterbox, July 2, 2010, http://www.useit.com/alertbox/ipad-kindle-reading.html (2010)
104. Nishino, H.: A 6dof fiducial tracking method based on topological region adjacency and angle information for tangible interaction. In: Proceedings of the fourth international conference on Tangible, embedded, and embodied interaction, TEI '10, pp. 253–256. ACM, New York, NY, USA (2010)
105. Nomura, S., Hutchins, E., Holder, B.E.: The uses of paper in commercial airline flight operations. In: CSCW '06: Proceedings of the 2006 20th anniversary conference on Computer supported cooperative work, pp. 249–258. ACM, New York, NY, USA (2006)
106. Norman, D.A., Nielsen, J.: Gestural interfaces: a step backward in usability. interactions **17**, 46–49 (2010)
107. Norrie, M.C., Signer, B., Weibel, N.: General framework for the rapid development of interactive paper applications. In: CoPADD 2006, Workshop on Collaborating over Paper and Digital Documents. Banff, Canada (2006)
108. Norrie, M.C., Signer, B., Weibel, N.: Print-n-link: Weaving the paper web. In: Proceedings of DocEng 2006, ACM Symposium on Document Engineering, pp. 34–43. Amsterdam, The Netherlands (2006)
109. Obendorf, H.: Simplifying annotation support for real-world-settings: a comparative study of active reading. In: HYPERTEXT '03: Proceedings of the fourteenth ACM conference on Hypertext and hypermedia, pp. 120–121. ACM Press, New York, NY, USA (2003)
110. O'Hara, K., Sellen, A.: A comparison of reading paper and on-line documents. In: CHI '97: Proceedings of the SIGCHI conference on Human factors in computing systems, pp. 335–342. ACM Press, New York, NY, USA (1997)
111. Olberding, S., Steimle, J.: Enabling erasing capabilities for anoto pens. In: Papercomp 2010 Workshop held in conjunction with UbiComp 2010 (2010)

References

112. Oviatt, S., Arthur, A., Cohen, J.: Quiet interfaces that help students think. In: UIST '06: Proceedings of the 19th annual ACM symposium on User interface software and technology, pp. 191–200. ACM Press, New York, NY, USA (2006)
113. Pauk, W.: How to Study in College. Houghton Mifflin (1989)
114. Pedersen, E.R., Sokoler, T., Nelson, L.: Paperbuttons: expanding a tangible user interface. In: DIS '00: Proceedings of the 3rd conference on Designing interactive systems, pp. 216–223. ACM, New York, NY, USA (2000)
115. Piolat, A., Olive, T., Kellogg, R.T.: Cognitive effort during note taking. Applied Cognitive Psychology **19**, 291–312 (2005)
116. Piper, A.M., Hollan, J.D.: Tabletop displays for small group study: affordances of paper and digital materials. In: CHI '09: Proceedings of the 27th international conference on Human factors in computing systems, pp. 1227–1236. ACM, New York, NY, USA (2009)
117. Piper, A.M., Weibel, N., Hollan, J.D.: Introducing multimodal paper-digital interfaces for speech-language therapy. In: Proceedings of the 12th international ACM SIGACCESS conference on Computers and accessibility, ASSETS '10, pp. 203–210. ACM, New York, NY, USA (2010)
118. Pittman, J.A.: Handwriting recognition: Tablet pc text input. Computer **40**, 49–54 (2007). DOI http://doi.ieeecomputersociety.org/10.1109/MC.2007.314
119. Plamondon, R., Srihari, S.N.: On-line and off-line handwriting recognition: A comprehensive survey. IEEE Transactions on Pattern Analysis and Machine Intelligence **22**(1), 63–84 (2000)
120. Polymer Vision: Rollable 6-inch SVGA display. http://www.youtube.com/user/PolymerVisionChannel
121. Probst, K., Seifried, T., Haller, M., Yasu, K., Sugimoto, M., Inami, M.: Move-it: interactive sticky notes actuated by shape memory alloys. In: Proceedings of the 2011 annual conference extended abstracts on Human factors in computing systems, CHI EA '11, pp. 1393–1398. ACM, New York, NY, USA (2011)
122. Rekimoto, J.: Pick-and-drop: a direct manipulation technique for multiple computer environments. In: UIST '97: Proceedings of the 10th annual ACM symposium on User interface software and technology, pp. 31–39. ACM, New York, NY, USA (1997)
123. Rekimoto, J., Saitoh, M.: Augmented surfaces: a spatially continuous work space for hybrid computing environments. In: CHI '99: Proceedings of the SIGCHI conference on Human factors in computing systems, pp. 378–385. ACM, New York, NY, USA (1999)
124. Robinson, P., Sheppard, D., Watts, R., Harding, R., Lay, S.: Animated paper documents. In: Proceedings of the 7th International Conference on human-computer interaction (HCI '97), pp. 655–658 (1997)
125. Rogers, Y., Ellis, J.: Distributed cognition: an alternative framework for analysing and explaining collaborative working. Journal of Information Technology **9**, 119–128 (1994)
126. Schreiber, D., Hartmann, M., Mühlhäuser, M.: MundoMonkey: Customizing Interaction with Web Applications in Interactive Spaces. In: ACM SIGCHI conference Engineering Interactive Computing Systems (EICS). ACM Press, http://www.acm.org/ (2009)
127. Schumacher, K., Liwicki, M., Dengel, A.: A paper-based technology for personal knowledge management. In: K. Hinkelmann, H. Wache (eds.) Wissensmanagement, *LNI*, vol. 145, pp. 289–298. GI (2009)
128. Schwesig, C., Poupyrev, I., Mori, E.: Gummi: a bendable computer. In: CHI '04: Proceedings of the SIGCHI conference on Human factors in computing systems, pp. 263–270. ACM, New York, NY, USA (2004)
129. Scott, S.D., Sheelagh, M., Carpendale, T., Inkpen, K.M.: Territoriality in collaborative tabletop workspaces. In: CSCW '04: Proceedings of the 2004 ACM conference on Computer supported cooperative work, pp. 294–303. ACM, New York, NY, USA (2004)
130. Seifried, T., Jervis, M., Haller, M., Masoodian, M., Villar, N.: Integration of virtual and real document organization. In: TEI '08: Proceedings of the 2nd international conference on Tangible and embedded interaction, pp. 81–88. ACM, New York, NY, USA (2008)
131. Sellen, A.J., Harper, R.H.: The Myth of the Paperless Office. MIT Press, Cambridge, MA, USA (2003)

132. Shaer, O., Leland, N., Calvillo-Gamez, E.H., Jacob, R.J.K.: The tac paradigm: specifying tangible user interfaces. Personal Ubiquitous Comput. **8**(5), 359–369 (2004)
133. Shneiderman, B.: Direct manipulation: A step beyond programming languages. Computer **16**(8), 57–69 (1983)
134. Signer, B.: Fundamental concepts for interactive paper and cross-media information spaces. Ph.D. thesis, ETH Zurich (2006)
135. Signer, B., Norrie, M.C.: Paperpoint: A paper-based presentation and interactive paper prototyping tool. In: Proceedings of TEI 2007, First International Conference on Tangible and Embedded Interaction, pp. 57–64 (2007)
136. Signer, B., Norrie, M.C.: Interactive paper: Past, present and future. In: Papercomp 2010 Workshop held in conjunction with UbiComp 2010 (2010)
137. Siio, I., Masui, T., Fukuchi, K.: Real-world interaction using the fieldmouse. In: UIST '99: Proceedings of the 12th annual ACM symposium on User interface software and technology, pp. 113–119. ACM, New York, NY, USA (1999)
138. Song, H., Grossman, T., Fitzmaurice, G., Guimbretière, F., Khan, A., Attar, R., Kurtenbach, G.: Penlight: Combining a mobile projector and a digital pen for dynamic visual overlay. In: ACM Conference on Human Factors in Computing (CHI'2009) (2009)
139. Song, H., Guimbretiere, F., Grossman, T., Fitzmaurice, G.: Mouselight: bimanual interactions on digital paper using a pen and a spatially-aware mobile projector. In: CHI '10: Proceedings of the 28th international conference on Human factors in computing systems, pp. 2451–2460. ACM, New York, NY, USA (2010)
140. Song, H., Guimbretière, F., Hu, C., Lipson, H.: Modelcraft: capturing freehand annotations and edits on physical 3d models. In: UIST '06: Proceedings of the 19th annual ACM symposium on User interface software and technology, pp. 13–22. ACM Press, New York, NY, USA (2006)
141. SONY: Rollable full color OLED display prototype. http://blog.sony.com/super-flexible-full-color-oled-display
142. Spindler, M., Stellmach, S., Dachselt, R.: Paperlens: advanced magic lens interaction above the tabletop. In: Proceedings of the ACM International Conference on Interactive Tabletops and Surfaces, ITS '09, pp. 69–76. ACM, New York, NY, USA (2009)
143. Steimle, J.: Integrating printed and digital documents: Interaction models and techniques for collaborative knowledge work. Ph.D. thesis, Technische Universität Darmstadt, Computer Science Department (2009)
144. Steimle, J., Brdiczka, O., Mühlhäuser, M.: CoScribe: Integrating Paper and Digital Documents for Collaborative Knowledge Work. IEEE Transactions on Learning Technologies **2**(3), 174–188 (2009)
145. Steimle, J., Gurevych, I., Mühlhäuser, M.: Notetaking in University Courses and its Implications for eLearning Systems. In: C. Eibl, J. Magenheim, S. Schubert, M. Wessner (eds.) DeLFI 2007: 5. e-Learning Fachtagung Informatik, pp. 45–56 (2007)
146. Steimle, J., Khalilbeigi, M., Mühlhäuser, M., Hollan, J.D.: Physical and digital media usage patterns on interactive tabletop surfaces. In: ACM International Conference on Interactive Tabletops and Surfaces, ITS '10, pp. 167–176. ACM, New York, NY, USA (2010)
147. Steimle, J., Weibel, N., Olberding, S., Mühlhäuser, M., Hollan, J.D.: PLink: Paper-based links for cross-media information spaces. In: CHI 2011 Extended Abstracts on Human Factors in Computing Systems. ACM Press (2011)
148. Stifelman, L., Arons, B., Schmandt, C.: The audio notebook: paper and pen interaction with structured speech. In: CHI '01: Proceedings of the SIGCHI conference on Human factors in computing systems, pp. 182–189. ACM Press, New York, NY, USA (2001)
149. Tabard, A., Mackay, W.E., Eastmond, E.: From individual to collaborative: the evolution of prism, a hybrid laboratory notebook. In: Proceedings of the 2008 ACM conference on Computer supported cooperative work, CSCW '08, pp. 569–578. ACM, New York, NY, USA (2008)
150. Taylor, S.: Txt-it notes: Paper based text messaging. Tech. Rep. MSR-TR-2008-172, Microsoft Research (2008)

151. Terrenghi, L., Kirk, D., Sellen, A., Izadi, S.: Affordances for manipulation of physical versus digital media on interactive surfaces. In: CHI '07: Proceedings of the SIGCHI conference on Human factors in computing systems, pp. 1157–1166. ACM, New York, NY, USA (2007)
152. Thakor, M.V., Borsuk, W., Kalamas, M.: Hotlists and web browsing behavior - an empirical investigation. Journal of Business Research **57**, 776–786 (2004)
153. Thayer, A., Lee, C.P., Hwang, L.H., Sales, H., Sen, P., Dalal, N.: The imposition and superimposition of digital reading technology: the academic potential of e-readers. In: Proceedings of the 2011 annual conference on Human factors in computing systems, CHI '11, pp. 2917–2926. ACM, New York, NY, USA (2011)
154. Tsandilas, T., Letondal, C., Mackay, W.E.: Musink: Composing music through augmented drawing. In: ACM Conference on Human Factors in Computing (CHI'2009), pp. 819–828 (2009)
155. Tsandilas, T., Mackay, W.E.: Knotty gestures: subtle traces to support interactive use of paper. In: Proceedings of the International Conference on Advanced Visual Interfaces, AVI '10, pp. 147–154. ACM, New York, NY, USA (2010)
156. Ullmer, B., Ishii, H.: The metadesk: models and prototypes for tangible user interfaces. In: UIST '97: Proceedings of the 10th annual ACM symposium on User interface software and technology, pp. 223–232. ACM, New York, NY, USA (1997)
157. Ullmer, B., Ishii, H.: mediablocks: tangible interfaces for online media. In: CHI '99: CHI '99 extended abstracts on Human factors in computing systems, pp. 31–32. ACM, New York, NY, USA (1999)
158. Ullmer, B., Ishii, H.: Emerging frameworks for tangible user interfaces. In: J.M. Carroll (ed.) Human-Computer Interaction in the New Millenium. Addison-Wesley (2001)
159. Underkoffler, J., Ishii, H.: Urp: a luminous-tangible workbench for urban planning and design. In: CHI '99: Proceedings of the SIGCHI conference on Human factors in computing systems, pp. 386–393. ACM Press, New York, NY, USA (1999)
160. Vernier, F., Nigay, L.: A framework for the combination and characterization of output modalities. In: Proceedings of DSV-IS2000, pp. 32–48. Springer (2000)
161. Volkamer, M., Vogt, R.: New generation of voting machines in germany – the hamburg way to verify correctness. In: Proceedings of the Frontiers in Electronic Elections (FEE 2006) Workshop (2006)
162. Wagner, D., Schmalstieg, D.: Artoolkitplus for pose tracking on mobile devices. In: Proceedings of 12th Computer Vision Winter Workshop (CVWW'07) (2007)
163. Watanabe, J.i., Mochizuki, A., Horry, Y.: Bookisheet: bendable device for browsing content using the metaphor of leafing through the pages. In: UbiComp '08: Proceedings of the 10th international conference on Ubiquitous computing, pp. 360–369. ACM, New York, NY, USA (2008)
164. Weibel, N., Fouse, A., Hutchins, E., Hollan, J.D.: Supporting an integrated paper-digital workflow for observational research. In: Proceedings of the 16th international conference on Intelligent user interfaces, IUI '11, pp. 257–266. ACM, New York, NY, USA (2011)
165. Weibel, N., Ispas, A., Signer, B., Norrie, M.C.: Paperproof: a paper-digital proof-editing system. In: CHI '08: CHI '08 extended abstracts on Human factors in computing systems, pp. 2349–2354. ACM, New York, NY, USA (2008)
166. Weibel, N., Signer, B., Norrie, M.C., Hofstetter, H., Jetter, H.C., Reiterer, H.: Papersketch: A paper-digital collaborative remote sketching tool. In: Proceedings of the International Conference on Intelligent User Interfaces (IUI'11), pp. 155–164. ACM Press (2011)
167. Wellner, P.: Interacting with paper on the digitaldesk. Communications of the ACM **36**(7), 87–96 (1993)
168. West, D., Quigley, A., Kay, J.: Memento: a digital-physical scrapbook for memory sharing. Personal Ubiquitous Computing **11**, 313–328 (2007)
169. Whittaker, S., Hirschberg, J.: The character, value, and management of personal paper archives. ACM Trans. Comput.-Hum. Interact. **8**(2), 150–170 (2001)
170. Wilcox, L.D., Schilit, B.N., Sawhney, N.: Dynomite: a dynamically organized ink and audio notebook. In: CHI '97: Proceedings of the SIGCHI conference on Human factors in computing systems, pp. 186–193. ACM Press, New York, NY, USA (1997)

171. Wilson, A.D.: Playanywhere: a compact interactive tabletop projection-vision system. In: UIST '05: Proceedings of the 18th annual ACM symposium on User interface software and technology, pp. 83–92. ACM, New York, NY, USA (2005)
172. Wilson, A.D.: Using a depth camera as a touch sensor. In: ACM International Conference on Interactive Tabletops and Surfaces, ITS '10, pp. 69–72. ACM, New York, NY, USA (2010)
173. Wilson, A.D., Benko, H.: Combining multiple depth cameras and projectors for interactions on, above and between surfaces. In: Proceedings of the 23rd annual ACM symposium on User interface software and technology, UIST '10, pp. 273–282. ACM, New York, NY, USA (2010)
174. Wimmer, R.: Grasp sensing for human-computer interaction. In: Proceedings of the fifth international conference on Tangible, embedded, and embodied interaction, TEI '11, pp. 221–228. ACM, New York, NY, USA (2011)
175. Wolfe, J.L.: Effects of annotations on student readers and writers. In: DL '00: Proceedings of the fifth ACM conference on Digital libraries, pp. 19–26. ACM Press, New York, NY, USA (2000)
176. Yeh, R., Liao, C., Klemmer, S., Guimbretière, F., Lee, B., Kakaradov, B., Stamberger, J., Paepcke, A.: Butterflynet: a mobile capture and access system for field biology research. In: CHI '06: Proceedings of the SIGCHI conference on Human Factors in computing systems, pp. 571–580. ACM Press, New York, NY, USA (2006)
177. Yeh, R.B., Brandt, J., Klemmer, S.R., Boli, J., Su, E., Paepcke, A.: Interactive gigapixel prints: Large paper interfaces for visual context, mobility, and collaboration. Tech. rep., Stanford University HCI Group (2006)
178. Yeh, R.B., Paepcke, A., Klemmer, S.R.: Iterative design and evaluation of an event architecture for pen-and-paper interfaces. In: UIST '08: Proceedings of the 21st annual ACM symposium on User interface software and technology, pp. 111–120. ACM, New York, NY, USA (2008)

Index

A-Book, 44, 63
Active reading, **6**, 16, 48, 91, 93, 103, 137, 145
Affordance, 3, 5, 13, 88
Android, 37
Annotation, 1, **5**, 12, 16, 39, 48, 50, 78, 93, 100, 103, 107, 116
Anoto pattern, **30**, 32, 95, 151
Anoto pen, 12, 23, **29**, 35, 42, 94, 109, 118, 129
ARTag, 22, 23
ARToolkit, 22
ARToolkitPlus, 23
Association area, 131, **134**, 157
Association gesture, 131, 157
Audio, 1, 11, 32, 41, 46, 70, 95, 133
Audio Notebook, 46
Augmented Surfaces, 33, **60**

Barcode, 21, 40, 41, 45, 135
Book, 1, 16, 41, 60, 96, 134, 135, 137, 170
Bookmark, 98, 151
Books with Voices, 41
Bush, Vannevar, 15, 91, 128, 131
ButterflyNet, 45, 81

Channel, 69, 89
Character error rate, 123
ChronoViz, 46
Client/server system, 101
Cognitive load, 14, 115
Cohabitation, 15
Collaboration, 4, 9, 14, 15, 17, 39, 47, 64, 70, 77, 79, 91, 99, 104, 108, 119, 120, 155, 175
  co-located, 99, 158
  remote, 99, 169
  sharing, *see* Sharing

Communication, 4, 9
Complementarity of information, 85
Computer aided design (CAD), 51
Conceptual activity, 77, 82
Context, 2, 97
Core interaction, **79**, 81, 83, 88
  altering shape, 80
  bridging, 80, 132, 133, 137, 152, 158
  clicking, 80
  combining, 80, 132, 137, 152
  inking, 80, 132, 133, 137, 152, 158
  moving, 80
  pen reorienting, 80
Cornell Notetaking Method, 6
CoScribe viewer, 96, 112, 134, 137
Cube, *see* Process cube

DataGlyphs, 23
DataMatrix, 22
Design guidelines, 88
Designers' Outpost, 61
Deskpad, 62
Dictionary, 125
Digital ink, 27, 65, 175
Digital Paper Bookmark, 98, 119, **151**, 162, 173
Digital pen, 10, 11, 13, 27, 39, 49, 70, 77, 94, 106, 129, 151
DigitalDesk, 26, 27, **58**, 81
Digitizing tablet, 28, 45, 46
Direct manipulation, 67, 172
Discontinuity
  of tools, 129
Display
  Anoto-enabled, 32, 96, 99, 129, 141, 142, 145, 146, 170
  pen-enabled, 6

Display quality, 4
Distributed Cognition, 75
Document, 134
  hybrid, 135, 139
  physical-only, 135
  printed-only, 139
  space, 93
Dynamic Shader Lamps, 33

E-book reader, 3, 9, 34
EAN, 21
Ease of learning, 115
Ease of use, 115
Ecological perspective, 15, 67, **74**, 94, 115, 168, 169, 176
Ecological view, 96, 98, 100, 134, 136, 137, **138**, 158, 161, 170
EdFest, 48
Electronic mail, 3, 9, 44, 46, 48, 53, 76
Electronic paper, 4, 10, 17, 34, 173, 175
EnhancedDesk, 59
Equivalence of information, 85

Facebook, 48
Feedback, 13, 89, 133
Fiducial, 21, 24, 60
Filter, 140
Flicker, 4
Fly Fusion Pentop Computer, 31
Folder, 98, 134, 136, 137, 139
Functional system, 75

Graphical User Interface, 13, 36, 63, 67, 69, 77, 80, 88, 108, 165, 167, 172
Greasemonkey, 135

Handwriting recognition, 16, 38, 43, 47, 65, 89, **122**, 126, 133, 157, 175
Hyperlink, 16, 41, 42, 44, 47, 53, 55, 58, 61, 78, 93, 96, 100, 127, 128, **131**, 140, 142, 143, 170
Hypertext, 41, 58, 131

Information ecology, *see* Ecological perspective, **76**
Information retrieval, 141
Interaction
  semantic level, 77, 83, 88
  syntactic level, 77, 83, 88
Interaction Lens, 44, 63
Interaction model, 67, 168
Interaction primitive, 79
Interaction technique, 15
Interactive paper, 10, 37

Interview, 116, 120, 143
iPaper, 37

Java, 101
JavaScript, 135

Keyboard, 118, 119
Kinect, 174
Knowledge work, 1, 3, 4, 15, 74, 76, 93, 127, 141, 149, 167

Layout, 7
Learning, 94, 149
  group meeting, 94, 99
Lecture, 116, 120
Letras, 37
Levenshtein Distance, 123
Livescribe, 31, 38, 43, 46, 54, 64, 109, 172
Logical page, 106, 107

Marble Answering Machine, 72
MediaBlocks, 72
Memex, 15, 58, 128
Meta-cognition, 153, 159
Mobile phone, 11, 17, 24, 37, 49, 57, 64, 70, 172, 174
Mobility, 8, 12, 13, 20, 29, 39, 43, 46, 48, 49, 64, 96, 152, 174
  micro-mobility, 9
Model-View-Controller, 71, 135
Mouse, 119, 130
Mozilla Firefox, 135, 137
Multi-user view, 112, 120
Multiplexing, 88, 137
MundoCore, 38, 135

Navigation, 4, 5, 85, 120, 137
Near Field Communication (NFC), 24
Nelson, Ted, 91, 131
Newspaper, 3
Note, 5, 93
Notebook, 12, 39, 42, 60
Notetaking, 6, 16, 43

Occlusion, 61
OLED, 4, 32, 34, 173, 175
Optical character recognition (OCR), 19, 25, 47, 58
Optical motion capture, 23, 34
Organizing information, 7

Page area, 79
Paper Augmented Digital Documents, 50, 53
Paper form, 49

Index 189

Paperless office, 1, 3, 9
PaperLink, 55
PaperPDA, 46
PaperPoint, 52
PaperProof, 50
PaperToolkit, 36
PaperWindows, 33, 63
PapierCraft, 53, 56, 81
PDA, 41, 44, 70
PDF, 96
Pen input, 10, 27
Pen-and-Paper User Interface, 2, **11**, 14, **69**, 168
Penlet, 32, 38
PenLight, 56
Physical page, 106
Pile, 7, 23, 62
Post-desktop user interface, 69
Post-WIMP user interface, 69
PowerPoint, 96
PPUI, *see* Pen-and-Paper User Interface
Printer, 10
Privacy
    visibility, 109, 110, 119
Process cube, 160
Process knob, 162
Process structuring, 159
Projector, 11, 33, 39, 56, 62, 64, 70, 71, 86, 168, 172

QR code, 22, 41
Questionnaire, 116, 143
Quickies, 40

reacTIVision, 23
Reading, 4, 5, 93
Reference, 78, 133
Representation, 71, 86
Resolution, 4
RFID, 21, 24, 40, 41, 98, 161

Scalability, 114
Scanner, 10, 19, 20, 46
Scope, 79, 134
Search, 1, 124
    full-text, 122, 126, 141
Semantic category, 159, 161, 164
Shared Design Space, 60

Sharing, 1, 108, 138, 153, 158, 159, 169
SIFT features, 25
Single-user view, 112, 120
Skype, 48
Sticker, 47, 98, 151
Structuring, 1, 93, 101, 151, 155
    process, *see* Process structuring

Tablet computer, 3, 5, 6, 9, 37, 56, 129
Tabletop, 6, 8–10, 23, 34, 39, 58, 60–63, 70, 141, 142, 145
TAC paradigm, 72
Tag, 139, 140, **149**
Tag Button, **110**, 119
Tag Menu Card, **157**, 164
Tagging, 53, 78, 93, 100, 149, 164
    category, 149
    free, 149
Tangible tool, 160
Tangible User Interface, 2, 71, 95
Territoriality, 8
Touch input, 8, 26
TUI, *see* Tangible User Interface
Twitter, 48

Ubiquitous computing, 2, 37
Ubisense, 25
Ultrasonic pen, 27, 55
Unified pen-based interaction, **129**, 145
URP, 72
User-centered design, 68

Video, 1, 46, 95
Video Mosaic, 59
Viewer, *see* CoScribe viewer
Visualization, 15, 100, 154
VoodooSketch, 34

Wall display, 58
WebStickers, 40
Wikipedia, 142
WIMP, 67, 69
Word error rate, 123
Word processing, 3, 4
World Wide Web, 3, 6, 16, 17, 39, 40, 44, 45, 53, 55–58, 65, 86, 93, 96, 99, 127, 131, 135, 141, 142, 151, 170, 175